SOCIAL INFRASTRUCTURE: NEW YORK

STUDENTS AT THE MOUNTAIN,
COPENHAGEN

SOCIAL INFRASTRUCTURE: NEW YORK

THE EDWARD P. BASS DISTINGUISHED VISITING ARCHITECTURE FELLOWSHIP / YALE SCHOOL OF ARCHITECTURE

DOUGLAS DURST / BJARKE INGELS

EDITED BY NINA RAPPAPORT,
JAMES ANDRACHUK,
AND ANDREW BENNER

TRANSMITTER PARK PIER, EAST RIVER, NEW YORK,
WXY AND DONNA WALCAVAGE

Published by:
Yale School of Architecture
180 York Street
New Haven, Connecticut 06511
www.architecture.yale.edu

Distributed by:
ActarD
355 Lexington Avenue, 8th Floor
New York, NY 10017
www.actar-d.com

This book was made possible through the Edward
P. Bass Distinguished Visiting Architecture Fellowship fund of the Yale School of Architecture. It is the
eighth in a series of publications of the Bass fellowship published through the dean's office.

Editors:
Nina Rappaport, publications director;
James Andrachuk (M.Arch. '12); and Andrew
Benner (M.Arch. '03), critic in architecture

Copy Editor: David Delp

Design: MGMT. design, Brooklyn, New York
Cover: Image by Nicky Chang, Avi Forman,
and Marcus Addison Hooks

Library of Congress Control Number: 2015934198
ISBN 9781940291253

CONTENTS

EDWARD P. BASS DISTINGUISHED VISITING ARCHITECTURE FELLOWSHIP

In 2003, Edward P. Bass, a 1967 graduate of Yale College who studied at the Yale School of Architecture as a member of the class of 1972, endowed this fellowship to bring property developers to the school to lead advanced studios in collaboration with design faculty. Mr. Bass is an environmentalist who sponsored the Biosphere 2 development, in Oracle, Arizona, in 1991, and a developer responsible for the ongoing revitalization of the downtown portion of Fort Worth, Texas, where his Sundance Square, which combines restoration with new construction, has transformed a moribund urban core into a vibrant regional center. In all his work, Mr. Bass has been guided by the conviction that architecture is a socially engaged art operating at the intersection of grand visions and everyday realities.

The Bass fellowship ensures the school's curriculum recognizes the role of the property developer as an integral part of the design process. The fellowship brings developers to Yale to work side by side with educators and architecture students in the studio, situating the discussion about architecture within the wider discourse of contemporary practice. The first Bass studio, led by Gerald Hines and Louis I. Kahn Visiting Professor Stefan Behnisch, in 2005, is documented in *Poetry, Property, and Place* (2006). The second Bass studio, in 2006, which teamed Stuart Lipton with Saarinen Visiting Professor Sir Richard Rogers ('62), engineer Chris Wise, and architect Malcolm Smith ('97), is documented in *Future-Proofing* (2007). *The Human City* (2008) records the Yale studio collaboration of Roger Madelin and Bishop Visiting Professor Demetri Porphyrios. *Urban Integration: Bishopsgate Goods Yard* (2009) documents the studio led by Nick Johnson and the FAT architecture partnership. *Learning in Las Vegas* (2010) records the work of the studio led by Charles Atwood and architect David M. Schwarz. *Urban Intersections: São Paulo*, which documents the studio led by real estate developer Katherine Farley and adjunct professor Deborah Berke, was published in 2011, and, in 2014, the work of the studio led by Hong Kong developer Vincent Lo and Saarinen Visiting Professors Paul Katz, Jamie von Klemperer, and Forth Bagley of Kohn Pedersen Fox was presented in *Rethinking Chongqing: Mixed Use and Super Dense*. With this eighth book in the series, it is a pleasure to present the research and studio led by real estate developer Douglas Durst of the Durst Organization, New York City, and Bishop Visiting Professor of Architecture Bjarke Ingels of the Bjarke Ingels Group, or BIG, which is based in Copenhagen and has an office in New York City.

INTRODUCTION

TRANSMUTING INFRASTRUCTURE: A VISIONARY PRAGMATISM

NINA RAPPAPORT, JAMES ANDRACHUK,
AND ANDREW BENNER

Is a visionary pragmatism possible? Between January and May 2012, an advanced design studio conducted at the Yale School of Architecture pursued that question, focusing on metropolitan New York. The erosion of infrastructure in and around New York City, its physical deterioration as well as the dissipation of political will to publicly fund it are some of the practical matters the studio tackles. Additionally, it addressed the intense pressure to build both market-rate and subsidized housing despite a dearth of developable land. Could this nexus of concerns find a working frame within which private equity might be persuaded to produce a socially engaged and healthy urban public realm? Or, practically speaking, how might a project that prizes, as a community's lifeblood, increased shared space and greater connectivity successfully collaborate with the bottom-line realities and naked self-interest of the open market?

The studio proposed the inhabited bridge, which incorporates a variety of programs in addition to transportation, as a likely starting point for testing potential synergies between public infrastructure and private development. Two sites were selected for their relevance and potential: the Tappan Zee Bridge, which connects Tarrytown and Nyack, and the proposed 42nd Street Bridge, which would extend of the famed thoroughfare over the East River to Hunters Point, Queens. Students were challenged to replace the Tappan Zee and reimagine its adjacencies for a piece of the city that has been stripped of public transit and pedestrian access. The idea of employing the 42nd Street Bridge as a site springs from Raymond Hood's 1925 notion of extending several major New York City streets across the rivers to better stitch together the boroughs, an idea that has gained pertinence in light of the new Cornell University campus on Roosevelt Island and the robust development of the Queens waterfront.

The final program and extent of the development were left open-ended in the studio brief, so students had ample latitude to adjust to site, uncover hidden potentials, and calibrate the revenue-generating parameters to make a convincing economic case. At semester's end, each project discovered unexpected alliances and symbiotic agendas. The Tappan Zee site demonstrated a capacity to address larger transportation concerns: linking regional rail spokes, coalescing food distribution from upstate, and providing a new terminus for the municipal water-taxi system, among other issues. Designing part of the span as a land bridge helped to manage the surplus of dredged soils and created new wetlands. For the 42nd Street site, one group of students linked the greenways on both sides of the East River, while another group diverted the bridge's path to connect to the new Cornell technology campus on Roosevelt Island and encourage a start-up culture that could expand into Hunters Point.

The studio was led by real estate developer Douglas Durst of the Durst Organization, a preeminent New York developer known for spearheading sustainable development in commercial and residential high-rises, and architects Bjarke Ingels and Thomas Christoffersen, partners in BIG, which has offices in Copenhagen and New York City. The Durst Organization and BIG are completing 625 West 57th Street, an innovative, 709-unit residential building near the Hudson River waterfront, in the Hell's Kitchen neighborhood of Manhattan.

The book begins with an interview with Bjarke Ingels, followed by an interview with Douglas Durst. Then, in an essay, Ingels discusses BIG's theoretical foundations, design approach, and important recent projects. Next, the four studio teams' research into the history of inhabited bridges, bridge typologies, New York City transport, and the physical, social, and economic conditions of the two sites is presented, followed by the teams' design work. Excerpts from the final review are included in this section and serve to illuminate the range of themes explored over the course of the studio.

The editors would like to acknowledge the work of the students who participated in the studio and whose cooperation was essential to this book: James Andrachuk, David Bench, Nicky Chang, Avi Forman, Tom Fryer, Marcus Addison Hooks, Bryan Kim, Karl Schmeck, James Sobczak, Susan Surface, and Craig Woehrle—all of the class of 2012.

We also extend our thanks to our copy editor, David Delp, as well as our graphic designers, Sarah Gephart and Pilar Torcal of MGMT. design, New York, for their elegant work.

625 WEST 57TH STREET (W57)

SCALING UP

CONVERSATIONS WITH DOUGLAS DURST AND BJARKE INGELS

BJARKE INGELS

Bjarke Ingels, founding partner of Bjarke Ingels Group, or BIG, discusses his ideas about practice and teaching.

NINA RAPPAPORT Architects must have a certain innate optimism to practice in a recession. How did you receive such large commissions during that time? How is your enterprising attitude received at home, and has it been a catalyst for other young firms in Denmark? Do you see architectural practice for young firms very different than in the United States?

BJARKE INGELS Eleven years ago, when the office was called PLOT, there hadn't been a new start-up office in Denmark for ten or twenty years. The general understanding was that it was impossible, which seemed like a self-fulfilling prophecy. And, of course, in Denmark, you qualify for work by having already done it, so it is a real catch-22. The way we broke the mold was by winning a handful of open international competitions. That gave us a voice and an opportunity to actually take on real challenges. Now, there is a whole forest of new Danish start-ups because the prerequisite for making it as a

new office is to just start. I think our example probably made a lot of people consider starting seriously. And, second, as our projects started getting built, it was increasingly clear that things could be achieved by hiring a young office with a different level of energy and a different approach than that of established practices. Perhaps New York projects have been entrusted to large corporate offices because those projects tend to be very, very large.

NR You have been able to break that trajectory by working with Douglas Durst. Do you think he saw a particular potential in your work that encouraged him to risk hiring someone without a track record in New York, considering all his projects are so local?

BI When I met Durst at a lecture in Copenhagen, we were set to build the 8 House, a 600,000-square-foot-building that would be the largest in the city. I did not think of him as a potential client in the beginning because it was clear that what he was doing was so different from what we were and vice versa. But we enjoyed interacting with him, and he came to our exhibit at the Storefront for art and architecture. Then, when I was teaching at Columbia, I invited him to participate in the studio. Later, he came to Denmark to visit his wife's family, and he came to our office, which was more like a courtesy visit. I think he was impressed with the work and the scale of the enterprise, and, shortly thereafter, he invited us to work on West 57th Street. I could imagine that this project would serve as an example not only for a different kind of architecture but for a different kind of architecture firm—perhaps a younger one.

NR Do you think your approach to designing the West 57th Street project poses new opportunities for apartment design in New York? And how is your working relationship with Mr. Durst compared to your relationships with the developers in Denmark? What has been your biggest challenge thus far?

BI In general, I was warned that NYC building regulations and NYC developers are the worst in the world. (Douglas even says that, even though there are sharks in the waters outside his home in West Palm Beach, he has no fear of swimming because, as a New York developer, the sharks show him professional courtesy.) In fact, all regulations are rigid and all developers are profit-oriented. Those are the rules of the game. But, even so, doing a 450-foot building is a hell of a lot easier in Manhattan than in Copenhagen!

One unique thing about Durst is that it is a third-generation family company. (Douglas likes to say that he is planning to retire to spend less time with his family.)

Therefore, they think long term. They are interested in quality—lasting attributes: energy efficiency, durability, and sustainability. They think beyond the presale of condos and much further into the future. And that makes them incredibly interesting to work with and for as an architect.

NR With twelve partners as well as project architects, how is your office organized? How do you divide the workload and responsibilities between your New York studio and the main office in Copenhagen?

BI Our CEO, Sheela Søgaard, is the only female partner and the only partner who is not an architect; she previously worked at McKinsey & Co. We hired a CFO to help us out, as well. Kai-Uwe Bergmann, who is an architect by training, does mostly business development, and we have five project leaders, including a design director in each of the two offices. I travel back and forth between New York and Denmark and oversee different projects of both offices, while my partners are in charge of the everyday reviewing. My involvement is quite intense in the first months and then, as things fall into place, as ideas crystallize and programs condense into architecture, my involvement becomes focused in the form of regular reviews.

NR What has been the trajectory of the organization?

BI If you divide it into two five-year chapters, the first period was PLOT, starting from scratch and building a body of work: then, BIG was building up a new identity and a more professional practice capable of taking on more comprehensive responsibilities. The end of that period was the formation of the partnership. In summer 2010, I distributed shares among the seven other partners. Later, we established the New York office with partner Thomas Christoffersen, who joined me from Copenhagen, and Beat Schenk, an old friend and Swiss architect who just joined the Partner group at BIG in the beginning of 2015.

NR What do you look forward to now that you have your 140-person office in New York?

W57 FROM THE HUDSON

BI The focus will be to balance out the offices over time so that both will be capable of doing intelligent, innovative, and relevant work, regardless of my involvement. Many offices struggle with their identity and integrity when the founding partners, eventually, move on. Only offices that have become cultures, or schools of knowledge, are capable of making that transition. It has been essentially an educational process, since half the partners have been my students and all the project leaders of the next generation have been students of the office. We have been educating each other so that it's not just a style, but a long-term perspective. With the development of the international firm, we have done better work than ever. It's a collective effort, even though there is a lot of individual contribution. I'm not saying that individuals don't matter, but it is all about the individual effort in the collective achievement. I am interested in creating the conditions that allow the individuals to blossom and prosper and evolve. I am in a fortunate position in that we have somehow been able to create a culture quite quickly—and I am personally into the idea of undermining the myth of the singular genius in favor of what you could call a cultural-sociopolitical movement.

NR You have compared the process of making buildings to storytelling and have even produced a book, *Yes Is More*. Why is narrative important for you not just in the office but also in terms of the public?

BI In architecture, more than anything else, how you get there is of great relevance because, in a way, a project is a snapshot of a fragment of society. People need to understand that a building often looks different because it performs differently. So, the behind-the-scenes stories are necessary for a full appreciation of architecture. Of course, as a user, you might appreciate it

COURTYARD

without understanding why. But a lot of people have opinions about buildings that they have never entered. They are disliked simply because they look different. In that sense, the back-story is a major part of the work of the architect because you can't just build buildings; you also have to persuade clients, collaborators, city officials, neighbors, opinion polls, and banks that something has to be built. Architecture doesn't have the luxury to prove itself by being built and then appreciated because, most likely a building will make it or break it before it even gets that far.

NR One of your design methods includes an interest in unexpected programmatic or social juxtapositions, and your work merges different forms into unusual hybrids. Does this approach help to create an identity for the project that you rely on as you go forward with a project—in convincing the client, for example? Some of your buildings take on the shape of a logo. Are those buildings basically the distillation of an idea into form?

BI I think it has to do with an economy of means. I am interested in complexity, which is different from complication. In computer programming, the shorter, more complex string of code makes the computer do the same function. It is a question of the density of attributes, essentially doing more with less. Therefore, we often distill our designs down to the simplest number of moves or the most blatant achievement of certain aspects; maximum effect with minimum means is also what gives it an iconic character. If you look at the evolution of company logos, they often start out as pictures and end up as emblems or symbols. In the evolution of writing, back in Mesopotamia, they started by drawing sheep and gradually invented something more symbolic, resulting in letters. This development from the pictorial to the symbolic is also a way of saturating meaning into the minimal amount of data.

NR But does such a building become a one-liner rather than a building that maintains the complexity that you discuss, like the Venturi "duck" or the scheme for your People's Building, in Shanghai, which happens to be in the shape of the Chinese character for "people"?

BI Of course, we have played a little bit once and a while, and sometimes we have even done it very blatantly, but the fact that it looks like the character was a coincidence. It was designed as a slab that opened up where it stood as a gate from the sea to the water. Once we discovered the coincidence, we embraced it. Quite often, there is a big difference between whether something is imposed or has evolved, whether it is generated through a process that has to do with the performance of the building or whether it is an image that is drawn. It is very rare—the only example I can think of is the Vejle Housing—that we design something like five buildings that look like the letters of the city name. Vejle had more to do with the fact that the main view of the city was from a highway bridge as it passes a fjord, and, as you look down over the city, the five buildings resemble a place card. Up close, they are beautiful, modern apartments, but, from the highway, they look like this amazing Hollywood sign.

NR Which you find amusing or playful...

BI In that case, it was simply irresistible. The three typologies were a V, two E's, and an L, which was also a J when mirrored. So, it was a series of well-known typologies, each of them different characters with really nice attributes.

NR In terms of your inheritance from Rem Koolhaas, for whom you worked on the Seattle Public Library, how do you think OMA's approach to program and culture plays out in your own work? What do you admire about his approach?

BI What I really liked about Rem was his almost journalistic approach; each project was not an artwork separate from the world but a specific architectural intervention in some economic, social, or cultural reality. I think where we probably differ is that, often, OMA's work is fueled by a critical approach, being against something, whereas, in our case, it is often affirmative. Nietzsche said that the affirmative forces always lose against the negative ones. We try to focus our interests and attention toward elements that we enjoy and accelerate or combine them with others in a straightforward way. The sorts of hybrids that emerge are products of unconventional, seemingly mutually exclusive sets of elements. So, whereas the revolutionary avant-garde has this need to go against something, leading to this Oedipal succession of father-murders, we are more focused on selecting and combining desirable elements in an almost evolutionary way to see what unexpected spin-offs—in a sense, children—emerge. I also think we might be a bit less formally restrictive than some of the other OMA offspring—we have fewer taboos, architecturally.

NR Formally speaking, how do you meet the design challenges of each project while maintaining your firm's identity? Eero Saarinen, for example, designed many different buildings, each with its own identity driven by a client's need—"the style for the job." Are you interested in an identifiable building style, or do you prefer to design according to each situation?

BI You don't need artifacts to have an identity if you already have a strong one. You don't need to hire an agency to give you a logo if what you do already says who you are. One way of projecting an identity is to limit your possibilities and modes of expression to a few categories. In that sense, although architects such as Zaha Hadid or Peter Eisenman are wildly expressive, they are also, in a sense, limited to doing "Zaha" or "Peter." However, we like to reserve the right to choose our weapons according to the case. Something that is ridiculously superficial in one situation might be right in another.

NR So, you are not making cookie-cutter buildings, even though they often exhibit similar characteristics.

BI I am not saying that the great artists and Pritzker Prize winners are doing cookie-cutter stuff, but the price you pay for having a strong identity that is rooted in a formal vocabulary is that it becomes a prison that restricts you. Zaha could never do the Glass House, for example.

NR Could you?

BI We could at least do a very classic 90-degrees-only project, as we are doing in Seoul—next to the "Cloud" towers by MVRDV—and, in the same breath, do the warped plane of the W57 project, in Manhattan, without any inherent contradiction or dilemma, but simply due to different conditions triggering different design decisions informing different vocabularies.

NR I am curious about how you engage social issues in your work, for example, architecture as a public art and how it impacts cities.

BI We engage social issues mostly as a general philosophy of inclusion. We try to design buildings that welcome people in various ways. Although a lot of the work we have done so far has been private, the 8 House expands the public realm into the building.

The public space we designed, called Super Park, is in the most ethnically diverse neighborhood in Denmark and includes extreme public participation. We invited citizens to nominate objects from their home countries to help create a vehicle with a sense of ownership and participation. It shows the diverse culture of Copenhagen to contradict the petrified image of Denmark as a homogenous culture.

NR Is the sprinkling of the space with artifacts genuine or rather gratuitous, like a Disneyland of cultures?

BI It's real. All of these elements are evidence of how the world is an ongoing global experiment in which people across the world have found ways of inhabiting urban space, of sitting together on a bench facing each other or away from each other. I was highlighting some of the behavior that already exists in this part of Copenhagen. We have Indians, Chinese, and other ethnic cultures existing right next to one another. So, rather than reducing the expression of the neighborhood to some cliché idea of Danishness, it is a more true expression of what Denmark is today.

NR It's tough to design for different cultures. Do you design a space as they would in their culture, or do you design your own space that they then occupy?

BI The idea of this space is to make it like a public playground, not an institutionalized collection of colored, veneered animals but a real place of discovery where there is a landscape of elements that provoke and stimulate different ways of interacting with the city and with each other. I think it is going to be an incredibly lively space.

NR What do you think the role of the architect is in city design?

BI As architects, our role is often reduced to the beautification of predetermined programs. A client calls us up on the phone—after having determined all issues of a project—and asks us to "make it nice." Architecture is society's physical manifestation on the crust of the Earth—an artificial part of the planet's geography. It is where we all live. Architecture is the stuff that surrounds us. And as architects constantly working in and with the city, you would think that we would be at the frontier of envisioning our urban future. However, while we sit at home waiting for the phone to ring or someone to announce a competition, the future is being decided by those with power—the politicians—or those with money—the developers.

I think it would be very beneficial for our profession—and our cities—if we, as architects, could become more proactive in injecting our ideas, our knowledge, and our expertise into the processes that precede design work. We could be better at pointing out untapped possibilities. In a way, our studio with Douglas Durst at Yale is an attempt to use academic exploration as a way to offer the city of New York alternative ways of solving infrastructural and urban challenges.

NR It is interesting the way that you have integrated public access into infrastructure for the new energy-waste plant in Copenhagen. Denmark is one of the few countries that have built a successful industrial symbiosis project, and it is fitting to create an inhabitable facility that integrates public space. How did you get the company to agree to do this? Was it your concept or part of the competition?

BI In the competition, essentially, they asked for a set design. If you go out of the norm, you have to be even more rigorous, professional, and calculated on budget and size than anyone else. Once the idea came up, it was irresistible. They were not lacking public space because there was a beach park, so a roof park didn't make much

sense. But by changing the topography, it would be one of the tallest hills in Denmark, at one hundred meters. So, we contacted the Danish Olympic Committee, the national alpine ski-team coaches, and the Alpine Skiing Association for feedback as to how it could operate as a sports training facility and as a business model. We delivered these supportive statements with the competition material so that the jury could see it was a practical idea. It was a pragmatic utopia to realize a dream in a small fragment, as a practical objective.

NR It is also the complex but simple evolution of a hybrid typology—energy plant and public ski resort.

BI It originated from a widely published ski resort in Norway. The roof was a navigable part of the terrain. In December, we received a commission for a ski resort in Lapland that resulted from this idea. We have a commitment to certain ideas that are not just one-liners but are part of a bigger philosophy. We are more empowered as architects every time we do a project, making the world a little bit more like we wish it could be. Essentially, as architects and as human beings, we have the means and the tools to realize our dreams, and every time we don't try we miss an opportunity.

Douglas Durst of the Durst Organization discusses recent projects and how the company operates.

NINA RAPPAPORT How is your New York City–focused company organized, and what is your philosophy regarding development? Do you have a mantra or some basic guiding principles?

DOUGLAS DURST We do have a protocol to follow. When we have issues or problems with any development project, the first response is to not panic. We analyze everything very carefully, and if we can't come up with a solution, then we go to stage two: we lower our standards. If that doesn't solve our problem, we go to stage three: we have a scapegoat for each project—usually our attorney, whom we blame for the problem and move on. As one reporter said, we have strong but flexible standards. Our philosophy is that each building has different goals and requirements. So, as the leaders, my cousin Jody and I learn from what we did in the past to see if we can improve the next time. In our parents' generation, they tended to construct each building in the same way as the previous one. That's the easiest way to build because you know your mistakes and you learn to live with them. We try to make new mistakes. We also try to make each building the best one we can, rather than making it the same as the last. We spend a tremendous amount of time studying materials and systems. Most people think, well, you are going to build a residential or commercial building, so you hire the builder and the architect, stir, and two years later you have a building. And there are some people who do do that.

NR How do you organize your teams and build collaborations with each project?

DD We have retreats out of the office to discuss potential problems. After dinner, we continue the discussions over drinks so that people are a little more relaxed. When I started in the business, the purpose of meetings was often to find somebody to blame for what was going on and why things weren't happening. For the first project I really worked on, 1155 Sixth Avenue, there were weekly meetings. About three-quarters of each meeting was spent with people pointing fingers as to why things weren't getting approved. The architect would blame the contractor, and the contractor would blame the engineer, and the engineer would blame the owner, and it would just go around in circles. Jody and I had gone through that, and we just weren't going to allow that to happen on our projects.

NR When do you bring an architect into a project discussion?

DD Almost immediately. A lot of my peers don't bring the architect in until later on and then have the architect work on spec. We don't believe in having an architect spec his time because we want to get the very best results for the building. The idea for Four Times Square was born sometime in fall 1995, and, as soon as it occurred to me that we could build a building there, I brought in Bob Fox and Bruce Fowle. We talked not just about the site but what would happen if we developed the entire block.

ONE BRYANT PARK

Social Infrastructure: New York

NR Was this a fruitful, dynamic collaboration?

DD It was the first time Jody and I had real oversight on a project, and it was Bob Fox who suggested the idea of retreats. Since we are very private and don't like getting up in front of a lot of people, it was not something we were interested in doing. It is still something we don't like to do, but we have found it to be so helpful in getting people to work together.

NR How does your experience with Four Times Square compare to that with One Bryant Park in terms of sustainability?

DD Four Times Square was the first large-scale office high-rise to be constructed as an environmentally responsible building. So, we were creating a new type of building. It was very exciting, but, naturally, some things did not work out, such as fuel cells, and others we did not consider, such as capturing rainwater, which we are doing here at Bryant Park.

NR Is the photovoltaic system at Four Times Square functioning and economical?

DD That was a real experiment. They have a payback of about twenty-five years and a life expectancy of about twenty, so it wasn't really an economic decision. We wanted to further the industry. The man who made the panels produced them in his garage, so we had to buy all the equipment in order to ensure delivery. We actually had to buy two sets of panels because it was not clear whether he was going to make them in time to finish the building. But he did. They produce power, but it is a fight with Con Edison to get them turned on.

NR What were the lessons learned?

DD Our main focus at Four Times Square was energy. We now realize that, while energy is important, the real issue is making the building as healthy and efficient as possible for the occupants. To bring in more outside air, it takes more energy to turn the fans and to temper and clean the air. If you are just looking at energy efficiency, you are not getting the effect that we think you should.

At Bryant Park, we paid more attention to water savings and preventing sewage-system overflow by capturing all the rainwater and reusing the groundwater. There is a lot of groundwater coming into the building, and the typical response used to be just to dump it into the sewer. We use it for flushing the toilets and in the cooling tower. At Four Times Square, we had a fuel cell, which has many applications, but it is not applicable to an office building. At Bryant Park, we have a five-megawatt cogeneration plant that produces about 80 percent of the power used in the building, and the waste heat is used to heat and cool the building. At night, when the building has low demand, the power is used to make ice, which cools the building during the day.

NR How has your perspective changed about buildings as living systems?

DD I see them as being more efficient and able to make better use of available resources, such as groundwater, natural gas to generate electricity, and natural light. These fixtures shut down during daylight hours. At Four Times Square, we looked at using fewer natural resources. We insisted that contractors recycle their own material, and they complained because of cost but actually found out that there were savings. Now, people don't even question it.

NR How are you involved in reevaluations and potential improvements to the LEED regulations?

THE HELENA, 601 WEST
57TH STREET, FXFOWLE,
NEW YORK, 2012

than it does across the river or in other parts of the city.

NR How is your firm involved in the World Trade Center site?

DD We are an adviser to the Port Authority of New York and New Jersey on finishing and tenanting the building. I was not in favor of all the office space being built down there—and I still think it could have been approached differently and completed over a longer time period—but that is behind us now. We have commitments from tenants for more than half the building, taking us to 2015. So, we believe it is going to be extremely successful.

NR Your next risk is with BIG on the residential project at 625 West 57th Street and the West Side Highway in New York City. I heard that you met Bjarke Ingels at a conference, and it was love at first sight.

DD I have been very vocal in complaining about LEED, but it has gotten people to think. It is a valuable resource, even though it is very expensive to adhere to. It is also somewhat subjective, but we don't have a better standard. I think, at some point, they are going to have to reevaluate the whole system, but that's a way off.

DD My wife is Danish. Six years ago, I was invited to give a talk about green buildings to the Copenhagen City Council. Europe has been way ahead of us in terms of energy efficiency but not in terms of total building efficiency. Bjarke is young and was, of course, even younger then. Toward the end of my talk, he asked, "Why do your buildings look like buildings?" (He now says he never asked that.) The question intrigued me, so I got to know him. For our fortieth anniversary, we went to Denmark and visited his office, and I was overwhelmed by the projects he was doing, so I talked to him about ours.

NR Have you taken different kinds of risk in light of the financial downturn? How has your business changed?

DD You have to take bigger risks because the banks require more equity. We haven't seen the decrease in land costs that would enable more projects to go forward. So, although construction costs have decreased considerably, New York is still not competitive with other markets. And it costs three times more to build in Manhattan

NR Is your working relationship different than it has been with other architects?

DD It has been a terrific collaboration. When we have to make changes for codes or economic

1133 AVENUE OF THE AMERICAS, EMERY ROTH & SONS, NEW YORK, 1970

reasons, we don't get a big pushback. Bjarke sees a problem and is very quick to find solutions. I have been very impressed with their grasp of the zoning here. They build all over the world, so I know they are very good at understanding different zoning and construction requirements in all the cities they work in.

NR Did the building's triangular shape around an open courtyard evolve from zoning, light, and air requirements or purely from design concerns?

DD We knew we wanted to have an interior courtyard on the European model, but the zoning made the building too tall—it would have been dark all the time. When we first started, the design had two high corners, like two towers, which didn't feel right. The site is zoned as commercial now, so we have to get that changed. We were going to show Amanda Burden, when she was head of city planning, what we were going to do. Bjarke and I decided that we would need to move the bulk to the East Side in such a way as to not impact the residential building that we had developed, the Helena. One problem was that 40 percent of the apartments were on the highway; Bjarke thought about it over the weekend and came up with the design. One of the amazing things about Bjarke is that he is willing to take a completely fresh look ever after he has finished with a design. Most architects on his level would say, "I have given you the design, and this is what it is." I am sure he wasn't thrilled about changing it, but I think he is very happy with the result.

NR What did you and Bjarke teach to the architecture students at Yale?

DD Just what we have been discussing: that it is a collaborative effort. There is no one person who takes the lead. You can design as nice a

building as you want, but unless the structure holds it up, the air-conditioning works, and the elevators work, it doesn't mean anything. The best architects we have worked with are those who think about a building from the inside out and make sure it functions for its intended use and isn't just built around a "design."

NR What about the financial side of development? Do you think architects need to know that, as well?

DD Yes, at least as far as the choice of materials and the operational costs of a building. Some buildings we are working on now are designed without any thought about how they are going to operate. That is fine for the architects because, once the building is done, they are gone. But we construct and manage our own buildings, so we have to be aware of how they are going to operate and what costs are created as a result of the design.

NR You are involved in many civic activities in New York City. How do you see your work as a developer contributing to the city's quality of life?

DD It is very important to us that we improve the urban quality of life. For example, we operate the New York Water Taxi, and I have been very active in the Hudson River Park. We believe design is very important, too. The lights on the Bryant Park spire weren't part of the original design, but people comment on what it does to the skyline at night. We are always trying to improve life in the city.

AERIAL VIEW OF 8 HOUSE

SOCIAL INFRASTRUC- TURE:

DENMARK AND AMERICA

BJARKE INGELS

Introduction

The way we work in our office is to spend time at the beginning of a project analyzing all the possible issues that could inform our design decisions: what are the key parameters, the crucial criteria, the biggest problems, and the greatest potentials? In answering these questions, we create projects that not only look distinct but also perform differently. This knowledge-driven design approach adds value

2009 UNITED NATIONS CLIMATE CHANGE CONFER- ENCE, COPENHAGEN

by synthesizing formal and programmatic ingredients. The goal of this approach is to expand architecture's role in the world in order to tackle the important issues we face today.

In Copenhagen, at the U.N. Climate Change Conference, in 2009, the grim faces of the world leaders said it all: the conference was a complete failure—none of the goals was reached. This failure of U.N. resolutions on

DANISH PAVILION AT EXPO 2010, SHANGHAI

MASSING DIAGRAMS
FOR 8 HOUSE

climate change comes out of a basic misconception in the general discussion of sustainability that compromises society's ability to come to a solution: our current quality of life has to be sacrificed in order to afford to be sustainable. This misconception is the main obstacle to proposing solutions to the world's most pressing issues. As an antidote, we wondered what if the opposite were true: what if sustainability increased our enjoyment of cities? And how can we make that happen?

All of our work is concerned with social or cultural purpose, and, increasingly, we are looking at how such considerations can tie into larger pieces of infrastructure and the networks they comprise. Cities and buildings are inhabitable landscapes; they are human-created ecosystems. The typical infrastructure of a city is necessary but comes at a cost: it takes up areas that are useful for one function but are, otherwise, hostile places for the people who live there. As an alternative, we believe we need to trace the flow through our cities of both resources and people. So, the question becomes, how can you increase the functionality of these infrastructures, and how can you make places desirable that are so hostile?

8 House

8 House (8 Tallet), completed in 2010, is a good example of our approach. It is on a new subway line in a suburban neighborhood that has as much density as the center, but it feels as if it were at the edge of civilization. If you are constructing a large building in the "middle of nowhere," how do you give it the character and identity and diversity that exist in a historical neighborhood?

We started by looking at the program of the building: shops and offices like to be closer to customers on the ground, and homes above. Because residential floor plates are shallower, it is possible to create space for a path to wind its way up the building, almost like a mountain path, allowing the apartments to open up as if they were townhouses. There are front yards enclosed by low walls, and the parapet holds the streetlights.

A shortcut had to go through the building, according to the master plan, so we pulled the building in at the middle. We stacked the amenities on top of each other, creating a vertical atrium with a staircase that connects everything from the ground to the roof deck, where there is a view high over the flat landscape. After optimizing the volume to maximize daylight and views, this big city block—seven hundred feet long by three hundred feet wide—suddenly became a three-dimensional urban condition.

FAÇADE OF 8 HOUSE

The 8 House is not just about a good-looking façade or an interesting form. This version of architectural alchemy expanded the possibility for social encounters: residents can meet their neighbors within the three-dimensional realm of the block—those meetings traditionally occur only at street level. The form of the building and the spaces within the form become a circulation landscape that allows for productive human interaction and possibility—a social infrastructure.

Super Park

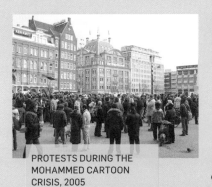

PROTESTS DURING THE MOHAMMED CARTOON CRISIS, 2005

We always try to make our projects generous to the public. In the case of Super Park (Superkilen), we took this attitude to the extreme, largely surrendering authorship. An opportunity appeared soon after the Mohammed cartoon crisis, in 2005: Many people perceived the publication of cartoons of the Prophet Mohammed by the Danish newspaper *Jyllands-Posten* as a public display of disrespect. It provoked criticism, outrage, and legal action from Islamic groups and, specifically, Danish Muslims. Around the same time, we were invited by the city of Copenhagen to create a new urban space in the neighborhood of Mjølnerparken, the most ethnically diverse in the country. In light of the crisis and, moreover, the fact that people from more than sixty different countries live in the neighborhood, it was clear to us the park had to create a sense of ownership, belonging, and integration.

Working with landscape architects Topotek 1 and the artist group Superflex, we proposed a simple overall plan for the park: a one-kilometer-long urban space wedging through the neighborhood in the tracks of the former rail yards. The plan consists of the "Red Square," where everything is in shades of red, including the bicycle lanes; the "Black Market," where everything is in black and white; and the "Green Park," where everything

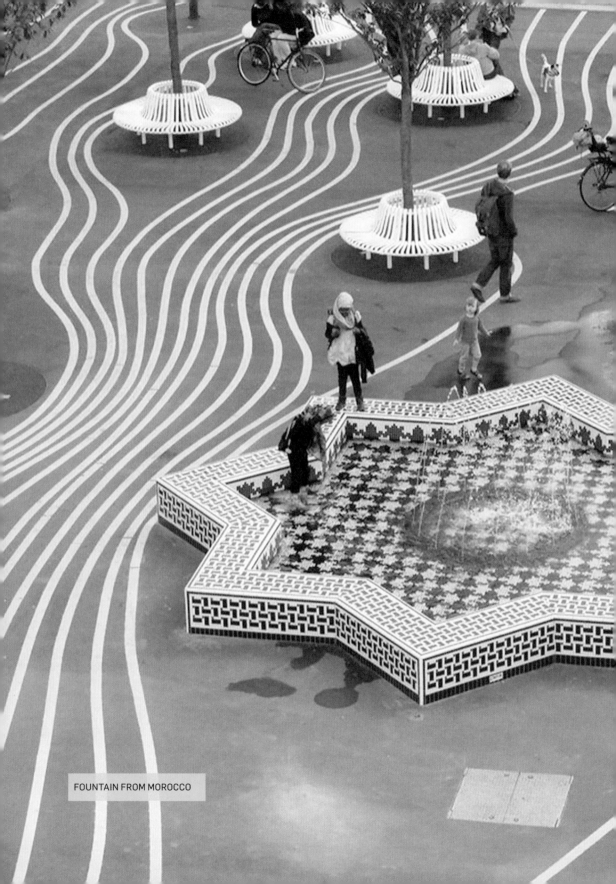

FOUNTAIN FROM MOROCCO

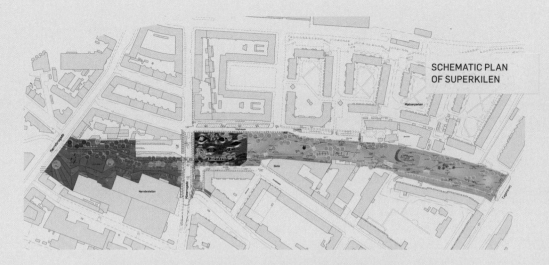

is in shades of green. The crucial element in the design, however, is the outdoor furniture. We asked, why not take advantage of the diversity of the neighborhood? Why not ask people to nominate outdoor public furniture from their home countries that could contribute to the park? We wanted to find the best of the best, and where better to do that than here?

The Danes do not necessarily make the nicest bench or the nicest trash bin. We do not eat Chinese food or Indian food to be nice—we eat it because we crave it, and the same goes for the furniture. So, in the Black Market, we installed a Moroccan fountain, because the Moroccans have an amazing tradition of architectural water features. We found an outdoor sound system from Jamaica, where residents can plug in iPods and put on spontaneous concerts—while obeying the local noise ordinances, of course. There is a boxing arena from Thailand, swings from Iraq, litter boxes from England, bollards from Ghana, bicycle racks from Finland, a bus stop from Kazakhstan, and manhole covers from France, Switzerland, and Israel—130 different objects in total. We even found hardy palm trees from China that can grow in the snowy climate of Copenhagen. Superkilen was a crowd-sourced urban design. As an urban space, it really reflects the contemporary cultural diversity of Copenhagen, rather than perpetuating a petrified perception of Danishness.

Copenhagen Waste-to-Energy Plant

Ninety-seven percent of homes in Copenhagen have district heating that comes from excess energy from power production. Seen as a resource, a ton of trash can produce as much energy as the oil in a barrel and two-thirds. Unfortunately, the waste-to-energy plant that converts the trash has to be located where the waste is generated and the energy and heat are consumed. Typically, these are ugly buildings, and they are not

VIEW OF EXISTING
PLANT FROM THE
CITY CENTER

welcomed. In light of this, how can the new power plant be a positive presence in the neighborhood?

The new Amager Bakke Waste-to-Energy Plant will be the largest building in Copenhagen, northeast of the center of the city, next to a public arena, and in an area where locals go water-skiing. It will be the cleanest plant of its kind in the world, treating 400,000 metric tons of waste annually and supplying 50,000 households with electricity. The chimney exhaust will be essentially nontoxic, containing only steam and some CO_2. Having situated the power plant, which is usually located in an off-limits industrial area, in close proximity to the city center, we thought we could capitalize on the project's location and environmen-

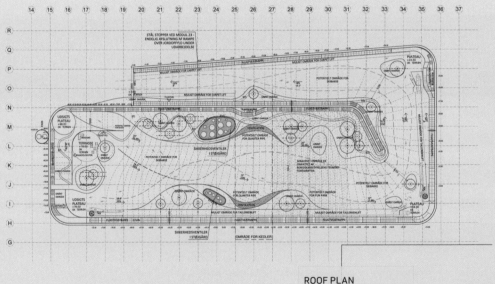

ROOF PLAN

tally friendly character as well as the necessary size of the building to provide a desirable public amenity to the residents of Copenhagen. Danes love to ski but have no mountains; residents of Copenhagen, for example, travel for hours to the south of Sweden to ski. Because of the footprint and height of this building, we can replicate two-thirds of the famous Swedish ski resort of Branæs on the roof of the building.

The way the doppelgänger resort is organized on the inside, the machinery is smaller at the ground, so the roof can be sloped, and the structure is wrapped in a continuous façade. The whole surface of the slope

VIEW OF THE NEW PLANT FROM A DISTANCE

VIEW OF THE SLOPE

VIEW OF THE FAÇADE
FROM THE INTERIOR OF
THE POWER PLANT

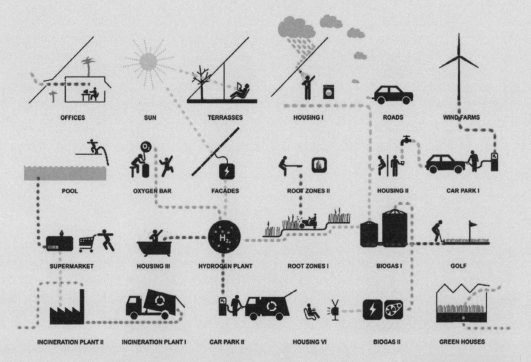

OFFICES   SUN   TERRASSES   HOUSING I   ROADS   WIND FARMS

POOL   OXYGEN BAR   FACADES   ROOT ZONES II   HOUSING II   CAR PARK I

SUPERMARKET   HOUSING III   HYDROGEN PLANT   ROOT ZONES I   BIOGAS I   GOLF

INCINERATION PLANT II   INCINERATION PLANT I   CAR PARK II   HOUSING VI   BIOGAS II   GREEN HOUSES

CONCEPTUAL DIAGRAM OF THE HUMAN-MADE ECOSYSTEM

is publicly accessible, so visitors can enjoy the view, even if they are not skiing. The power plant's ventilation elements are incorporated as landscape features.

Trees are integrated into the slope itself, and the façade has planters made of prefabricated aluminum bricks, which filter natural light into the interior, integrating structure and aesthetics. We worked with consultants in Germany and Denmark to develop a special type of chimney that accumulates steam and releases it in the form of a giant smoke ring. We consider this the ultimate artistic expression of hedonistic sustainability. A ton of CO_2 is hard to picture, but if you come to Copenhagen and count five smoke rings, that is a ton of CO_2. One of the main factors of behavioral change is knowledge: if people do not know, they cannot act. By projecting the carbon footprint of all the citizens of Copenhagen onto the city's skies, everyone will have a complete understanding of the impact of their actions.

The power-plant project is a material elaboration of the idea that cities and building are ecosystems. The plant harvests local resources— rainwater, sunlight, wind—to turn waste into a resource, thereby creating a symbiotic relationship with Copenhagen. At the same time, it acts as multi-use infrastructure, providing recreational and educational opportunities for the city dwellers.

COPENHAGEN HARBOR BATH EXTENSION

LINCOLN TUNNEL

34th STREET

BANK STREET

HARRISON STREET

BROOKLYN BRIDGE

MONTGOMERY STREET

LOUNGING
FOOD + DRINK
BOATING

BOATING

LOUNGING
SPORTS
BOATING
PLAYGROUND
LOUNGING
FOOD + DRINK
SPORTS
FERRY
BOATING
SPORTS

ENVIRONMENTAL EDUCATION

BIKING

SPORTS
LOUNGING
SPORTS

PLAN OF LOWER
MANHATTAN SHOW-
ING FLOODING "PINCH
POINTS" AND ADJACENT
NEIGHBORHOODS

FERRY
FOOD + DRINK
FARMING

FOOD + DRINK
LOUNGING
FERRY

New York Resiliency

We have given this idea of social infrastructure as ecosystem a new iteration as we look at New York City's resiliency challenges for Rebuild By Design. How do we prevent the effects of Hurricane Sandy, which devastated parts of the city in 2012, from happening again? How do we protect the areas in which the city interfaces with the water in a way that does not incarcerate the city behind walls of resiliency infrastructure? Through architecture and urban design, how can this project fulfill its federal-, state- and city-mandated benchmarks and legal obligations, yet offer enjoyment? The challenge is to make an infrastructure that comes prepackaged with programs and improvements for the community.

We see our proposal as the love child of Robert Moses and Jane Jacobs: Approaching the problem from the top down, we created a continuous line of defense; we then considered, from the bottom up, the existing piece of city and materialized the project so that, depending on where you are, the environment and views look and feel completely different. Historically, the growth of New York City since the seventeenth century has relied upon reclaimed areas, and these areas are the most flood-prone. Despite this, there are "pinch points" at regular intervals around the edge of the island where the water does not come very far inland. We created autonomous compartments, so one leak won't flood the whole city and the adjacent neighborhoods can extend toward the rivers.

Sandy hit the East Side of Manhattan particularly hard. This tract of the city, which is one of the densest in lower Manhattan, is primarily a residential neighborhood that comprises a wealth of cultures and ethnicities. A complex site, it is one of the last vestiges of affordable housing in

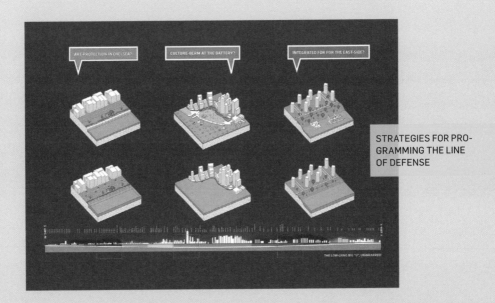

STRATEGIES FOR PROGRAMMING THE LINE OF DEFENSE

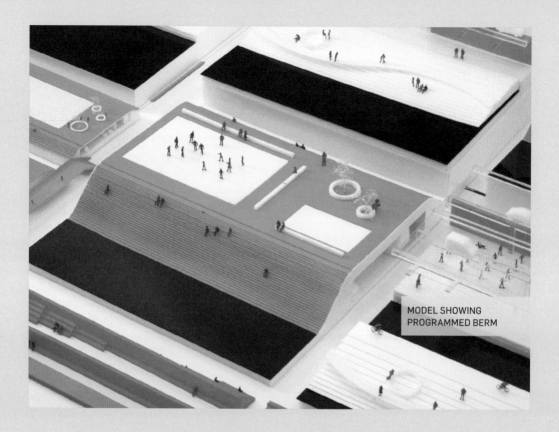

MODEL SHOWING
PROGRAMMED BERM

Manhattan. The neighborhood is underserved by the MTA and, like other, similar neighborhoods, chronically underfunded. The neighborhood, statistically, has open space—one fifth of Manhattan's total—but it is mostly paved and impervious and, thus, prone to flooding. And even though it is right next to the river, the East Side waterfront does not provide a pleasant, salubrious environment: it is tortured by infrastructure, especially Franklin D. Roosevelt Drive, making it hard to even access the water. Consequently, a thoughtful intervention would have a big, positive impact on a lot of different levels.

Forty years ago, it was the same situation on the West Side, which, since then, has been vastly improved and is now very desirable, thanks to significant investment. Looking ahead, we thought we could take some of the ideas from there—modified roadways, for example—and apply them to the East Side. During the design process, we tried to interface with the area's user groups and interest groups and spent time prototyping with them. The result is that the hundreds of millions of dollars the government will have invested in resiliency in the neighborhood will reflect what the residents want for themselves. Even though we are usually privately commissioned, public participation is of great importance. The sensitive context and competing interests of a public project greatly determine the design.

PROPOSED RIVER EDGE,
LOWER EAST SIDE

    The apartments on the vulnerable first floors of buildings would be turned into community amenities, and new buildings would be built to accommodate more affordable housing and consolidate the apartments. The highway could be partly submerged. Subway lines could be added. Naturally, there would be an elevated park along the water, above the flood line; the open space between the social housing would connect to it. Part of resilience infrastructure is to create a certain defense at a certain height, so we explored whether the barrier could incorporate a series of social programs. For instance, an amphitheater, a space for art projects, or landscape elements could be part of the flood defense. Further, the flood barrier could be incorporated in a space that is already lost in the form of landscape: a roadway median. In a flood event, only one lane would be sacrificed. The city sees great potential in using federal funding for resilience, using different financing models to improve the condition of a neighborhood that has fewer private resources than other neighborhoods. Designed to uplift a low-income, underserved neighborhood, our waterfront project is just one of many projects being financed by federal Flood Protection Act and American Recovery and Reinvestment Act funds and a variety of financing models, including public-private partnerships.

Conclusion

Design is informed by its physical and social situation. This attitude was promoted by Bernard Rudofsky, who invented the term "vernacular architecture" in his *Architecture without Architects* exhibition at the Museum of Modern Art, in 1964. An introduction to non-pedigreed architecture, the exhibit was a criticism of the International Style for its tendency to produce buildings that all looked the same, no matter the context. A major part of the International Style's aesthetic is determined by mechanical engineering and building services: each requirement gets a machine. Mechanical ventilation, electric lights, central heating, and air-conditioning—all these systems make users independent of the orientation of openings and the thickness of walls. As a result, the architecture no longer performs any functions entirely on its own. At the same time, the mechanical services are the compensation for the building being bad at what it was designed for: human occupation.

AERIAL VIEW OF PROPOSED RIVER EDGE, LOWER EAST SIDE

Rudofsky looked at examples of the vernacular, such as the white-cube buildings of Greece that reflect the harsh sunlight, Arctic igloos made from the only practically available building material, Chinese underground courtyard houses protected from prevailing winds, and, in Yemen, dense courtyard neighborhoods with ventilation chimneys that capture prevailing winds and create natural ventilation in a six-story-tall urban fabric. These architectural vocabularies do not emerge from aesthetics or academics but are evolved empirically using local materials to respond to local climates and local ways of life. Essentially, that idea is to make architecture that once again performs for people. We want to use the most sophisticated design processes to make buildings independent of their necessary mechanics and put back the technical and social performance attributes into the design.

OIA, SANTORINI, GREECE

STUDIO
BRIEF
GOALS AND
METHODS

The aim of the studio was to explore potential synergies between public infrastructure and private development. The brief proposed inhabited bridges at two sites: the Tappan Zee Bridge, which is located to the north of the city on the Hudson River, and the proposed 42nd Street Bridge, which would be an extension of the street to Hunters Point, Queens. Apart from helping to facilitate traffic, the bridges also were to contain various social programs, parks, pathways, and programs for residential, commercial, and cultural activities. The specific mix of programs at each site was left open to give the students leeway to adjust to the site and uncover innovative potentials through their research. The studio's intention was to see whether a public infrastructure investment could be significantly financed through private development and whether a purely utilitarian piece of infrastructural equipment could be imbued with social activities and used to create public space.

The semester kicked off with an introduction by Douglas Durst on the mechanics of public-private partnerships and the vicissitudes of the New York City approval process. Bjarke Ingels followed this with a seminar that elucidated his thoughts and experiences pursuing, in a variety of contexts, socially and environmentally engaged architecture. He emphasized the benefits of letting a research-based process guide and refine design directions and rhetorical strategies.

The students' work began with investigations into a number of topics: the history of inhabited bridges, which were collected into a database of precedents; bridge typologies and global construction trends, delving into the history and evolution of bridges in and around New York City; the physical, political, and economic organization of the New York City transit system; an analysis of the environmental and energy systems and demands in the city; site-specific studies of each location's history, condition, and stakeholders; a market analysis of real estate in communities directly adjacent to the sites and projections for likely real estate values.

Concurrent with this research, the studio took a trip to New York City to visit the sites, meet with city planners and the Durst Organization, and see relevant developments. Later, there was an excursion to Denmark, Sweden, and Norway to observe several successful housing, cultural, and infrastructural developments funded through public-private partnerships.

Upon their return from the Nordic trip, the students settled into four groups, with two groups focusing on one site and two groups focusing on the other. Another round of research delved into the specific problems and potentials posed by the sites. and the groups began to form arguments for how an insertion of housing might resolve or realign those possibilities. Leading into the midterm, each group was asked to craft for its project a case for a public-private partnership and then propose three options for implementation. This assignment required work at both a master-planning scale and at the scale of a more specific development of unit types. An emphasis was placed on establishing clear structural, circulation, and energy strategies, which, taken together, were the main design criteria of the midterm review.

Following the midterm reviews, the teams were asked to design a consolidated proposal by integrating the best aspects of the variants. Along with this process, a number of workshops were held with outside experts. A structural engineer helped refine strategies for the bridge construction and related building issues. Real estate and marketing matters were honed with the help of Nancy Packes, who is an esteemed real estate consultant. Douglas Durst and his team reviewed and shared their insights on strengthening the pro forma financial projections.

At the final review, the students gave the jurors their formal presentations, introducing the site, studio goals, and a summary of their research findings. Then, each group presented its final proposal, integrating their research, analysis, and design ideas. In their presentations, the students employed diagrams, which described the program, structure, energy, and circulation strategies; master-plan drawings and models; large-scale sectional drawings and models; study models; financial analyses and evocative perspectival imagery.

STUDENTS AT 8 HOUSE,
COPENHAGEN

TRAVEL

NEW YORK CITY, DENMARK, SWEDEN, AND NORWAY

STUDENTS ON THE NEW YORK WATER TAXI TOUR ALONG THE EAST RIVER

AT THE ROYAL DANISH ACADEMY OF FINE ARTS

In New York City, the students met with city planners to discuss the concept of using an inhabited bridge and how they might build on the initiatives of the city's long-term plan, PlaNYC. The Durst Organization then hosted the students at One Bryant Park, giving insight into the aspirations and design of their completed LEED Platinum tower as a model of sustainable urban development. A tour of the two sites by water taxi followed, along with a chance to see One World Trade Center, still under construction. The day culminated in a visit to the New York City offices of BIG, where the students met with an engineer from AECOM to talk about the design and challenges regarding the new Tappan Zee Bridge.

In February 2012, the studio's twelve students traveled to Denmark, Sweden, and Norway to observe successful infrastructural developments funded by public-private partnerships. In Copenhagen, they explored the city's transit systems, open spaces, and notable historical and recent works of architecture. While there, they presented preliminary research to several partners at BIG's Copenhagen office. They also visited the offices of accomplished bridge designer Dissing + Weitling Architects, in Copenhagen, and met with Camilla van Deurs at Gehl Architects, also in Copenhagen, to discuss the firm's contributions to reviving public space in New York City and other cities. During their time in the region, the students explored the emerging transnational loop that includes the Swedish cities of Malmö and Lund. Later, the studio flew to Oslo, where the students toured the offices of Oslo S. Utvikling AS (OSU), the development company behind Oslo's mixed-use waterfront Barcode project. They also saw the Statoil regional headquarters by the architecture firm A-Lab and paid a visit to the offices of Snøhetta, the architecture firm that designed Oslo's new waterfront opera house.

VIEW FROM 8 HOUSE

STUDENTS AT CENTRAL STATION, COPENHAGEN

ØRESTAD COLLEGE, COPENHAGEN, 3XN ARCHITECTS

STUDENTS WALKING THROUGH CONSTRUCTION SITE FOR THE DANISH NATIONAL MARITIME MUSEUM, HELSINGØR, BY BIG

ØRESUND BRIDGE, LINKING COPENHAGEN
AND MALMÖ, SWEDEN

COPENHAGEN CITY CENTER

HOLMENKOLLEN SKI JUMP, OSLO

WALKING THROUGH THE
BARCODE DEVELOPMENT
SITE, OSLO

SEMINAR AT GEHL ARCHI-
TECTS, COPENHAGEN

STATOIL HEADQUARTERS
CONSTRUCTION SITE, OSLO

OSLO OPERA HOUSE

BRIDGE HISTORY AND PRECEDENTS

INHABITED BRIDGES

The earliest examples of "programmed" bridges date back to the seventh century b.c.e., beginning with the first Roman aqueducts. The emergence of "habited" bridges—bridges whose primary purpose is neither water transportation nor energy generation—occurred during the Middle Ages, a period of increasing urban development and densification. A significant project during this time was the "Old" London Bridge, which was constructed during the twelfth century. With private commercial and residential programs above and water- and energy-related public infrastructure at the base, London Bridge, as vividly depicted in numerous medieval engravings, was the first truly mixed-use habited bridge. The following millennium witnessed countless variations on this theme. From the piers of Manhattan to the shores of the Pearl River, though some lie in ruins, these projects document architecture's continual fascination with the programmed bridge and testify to the form's enormous appeal.

CONSTRUCTION OF THE OLD LONDON BRIDGE WAS STARTED ABOUT 1175; BUILDING PLOTS ALONG ITS LENGTH WERE LEASED TO HELP RECOUP THE COSTS OF CONSTRUCTION. ULTIMATELY, DOZENS OF BUILDINGS WERE CONSTRUCTED ALONG THE BRIDGE, LEAVING THE ROADBED ALMOST ENTIRELY ENCLOSED

"*PONTE MAGNIFICO*," GIOVANNI BATTISTA PIRANESI, CA. 1750

PULTENEY BRIDGE, BATH, ENGLAND, 1769–1774

THE PONTE VECCHIO (INITIALLY CONSTRUCTED CA. 996 AND RECONSTRUCTED IN THE MID-1300S), IN FLORENCE, ITALY, SPANS THE RIVER ARNO AND WAS HOME TO A VARIETY OF MERCHANTS

SIOSE BRIDGE, ISFAHAN, IRAN. THIS DOUBLE-DECKED ARCHED STONE BRIDGE SPANS THE ZAYANDEH RIVER AND WAS CONSTRUCTED BY THE SAFAVID DYNASTY IN THE EARLY SEVENTEENTH CENTURY. THE HIGH WALLS ALONG THE ROADWAY WERE BUILT TO PROVIDE SHELTER FROM WINDS

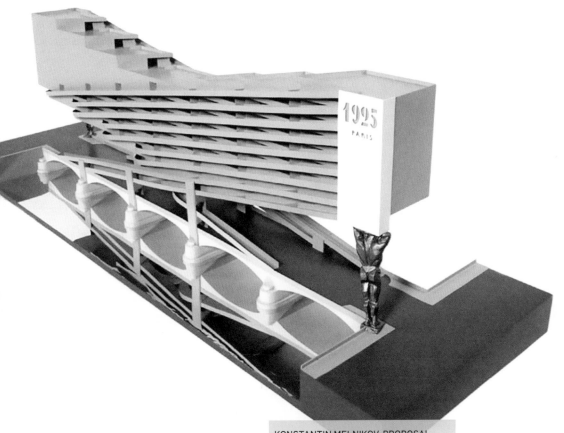

KONSTANTIN MELNIKOV, PROPOSAL
FOR A PARKING GARAGE OVER THE
SEINE RIVER, PARIS, 1925

OPENED IN 1912, CHENGYANG
BRIDGE, IN GUANGXI, CHINA,
COMBINES THE CHINESE PAVILION
TYPOLOGY WITH MULTIPLE OPEN
CORRIDORS SUPPORTED ON THREE
STONE PIERS

ZAHA HADID, BRIDGE OVER THE THAMES,
ENTRY TO PEABODY TRUST HABITABLE
BRIDGE COMPETITION, 1996

Bridge History and Precedents

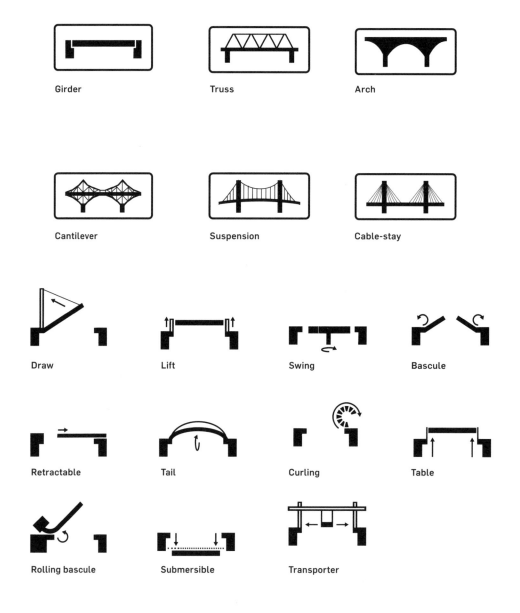

Girder

Truss

Arch

Cantilever

Suspension

Cable-stay

Draw

Lift

Swing

Bascule

Retractable

Tail

Curling

Table

Rolling bascule

Submersible

Transporter

THE BASIC BRIDGE TYPOLOGIES, THEIR VARIATIONS
AND COMBINATIONS

BRIDGE TYPOLOGIES

In order to better understand the embedded potential of medium- and long-span bridges, the studio's precedent research into existing bridge typologies began with an analysis of basic morphologies. The history of innovation in bridge engineering has resulted in the six basic types upon which all bridges are based: girder, truss, arch, cantilever, suspension, and cable-stay. Each type maximizes its structural capabilities in its own way. These basic types can be combined with one another into various hybrid typologies. This hybridization allows bridge designers to mitigate the weaknesses of any one type and solve spanning challenges that would be impossible otherwise. Additional elements, including movable components, augment the capabilities and forms of these bridges, allowing for even greater flexibility in their deployment. These bridge typologies provided inspiration for the students' early design explorations. Because of the studio's focus on real-world situations, built examples of each bridge type were analyzed in terms of span capabilities, construction methodology, cost, geographic location, and current industry trends. As studio projects began to develop, these precedents were used by the students to argue for the rationale and feasibility of both their design and architectural decisions.

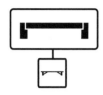

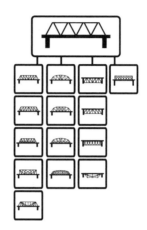

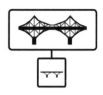

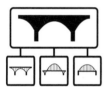

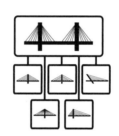

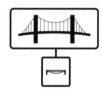

THE BASIC BRIDGE TYPOLOGIES, THEIR VARIATIONS
AND COMBINATIONS

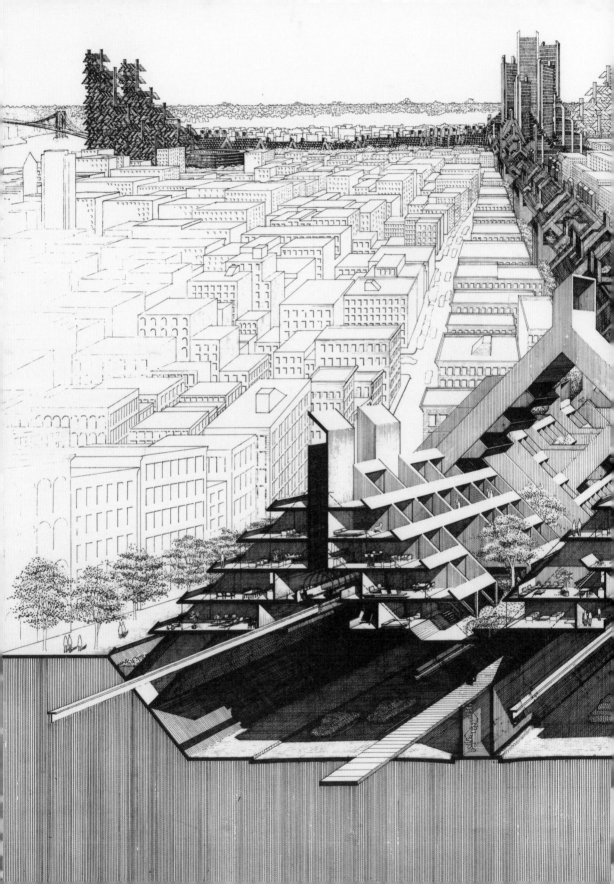

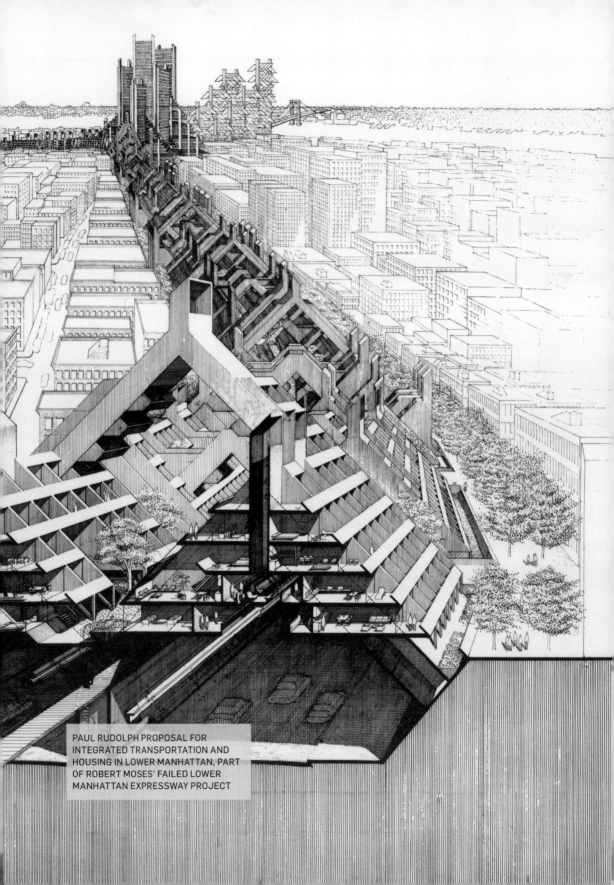

PAUL RUDOLPH PROPOSAL FOR INTEGRATED TRANSPORTATION AND HOUSING IN LOWER MANHATTAN, PART OF ROBERT MOSES' FAILED LOWER MANHATTAN EXPRESSWAY PROJECT

NEW YORK CITY BRIDGES AND TRANSPORTATION

New York City has a history of building exceptional bridges, each of which represents the era in which it was built. The earliest period of major bridge-building happened during the industrial era, from 1880 to 1920, and coincided with the amalgamation of the five boroughs into the City of New York, in 1898. These bridges were characterized by multimodal transportation and included road, pedestrian, and trolley connections; the 1883 Brooklyn Bridge, the 1903 Williamsburg Bridge, and the Manhattan and Queensborough bridges, both completed in 1909, are the major projects from this period. After the Great Depression and the emergence of Robert Moses and his Triborough Authority, bridge construction focused on automotive traffic in order to generate profits that could be fed back into bridge and highway construction—for example, the 1931 George Washington Bridge, the 1936 Triborough Bridge, and the 1939 Bronx Whitestone Bridge—as other modes of transportation fell by the wayside. Then, the scale of bridge construction increased, and access to bridges was further limited: the 1961 Throgs Neck Bridge and 1964 Verrazano-Narrows Bridge were restricted to automobiles and closely integrated with the highway system.

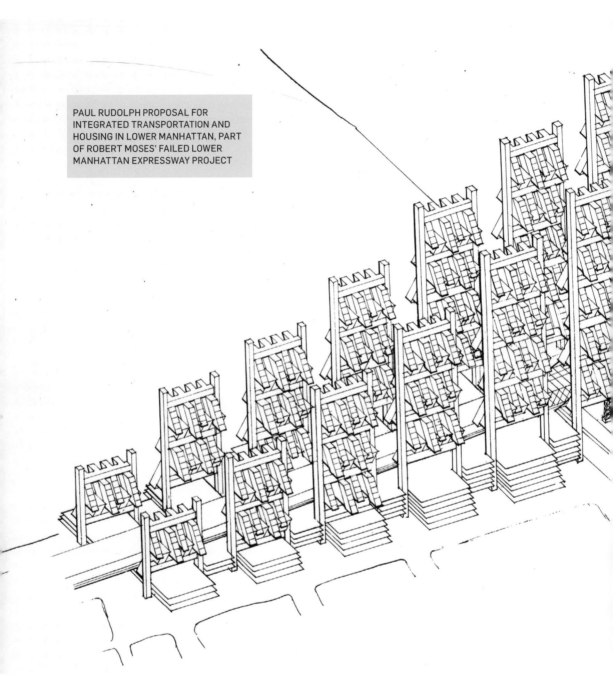

PAUL RUDOLPH PROPOSAL FOR
INTEGRATED TRANSPORTATION AND
HOUSING IN LOWER MANHATTAN, PART
OF ROBERT MOSES' FAILED LOWER
MANHATTAN EXPRESSWAY PROJECT

RAYMOND HOOD, "PROPOSAL FOR MANHATTAN, 1950," MONTAGE OF AERIAL PHOTOGRAPH AND DRAWING, 1929

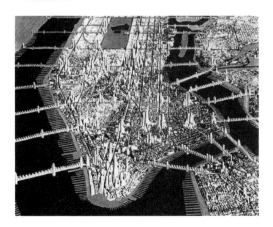

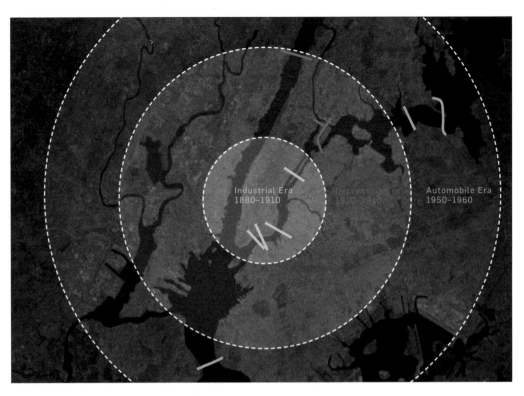

Industrial Era
1880–1910

Depression Era
1930–1940

Automobile Era
1950–1960

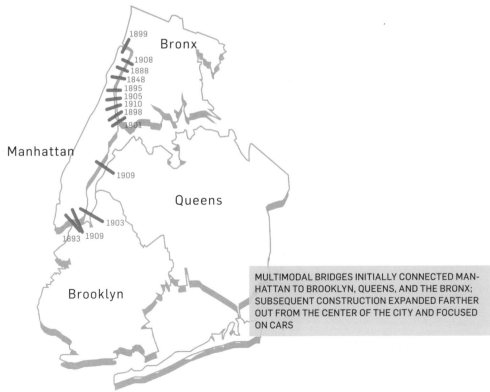

1899

Bronx

1908
1888
1848
1895
1905
1910
1898
1901

Manhattan

1909

Queens

1903
1893 1909

Brooklyn

MULTIMODAL BRIDGES INITIALLY CONNECTED MAN-
HATTAN TO BROOKLYN, QUEENS, AND THE BRONX;
SUBSEQUENT CONSTRUCTION EXPANDED FARTHER
OUT FROM THE CENTER OF THE CITY AND FOCUSED
ON CARS

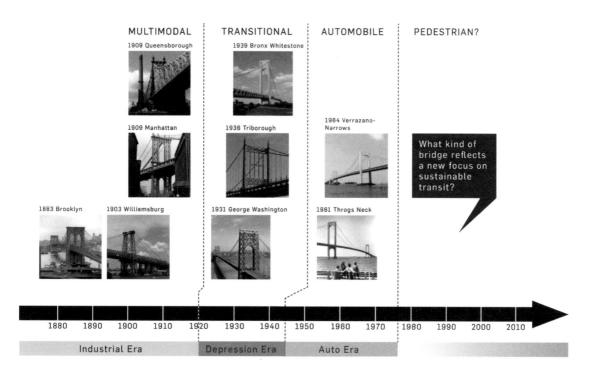

MULTIMODAL | TRANSITIONAL | AUTOMOBILE | PEDESTRIAN?

1909 Queensborough
1939 Bronx Whitestone
1964 Verrazano-Narrows

What kind of bridge reflects a new focus on sustainable transit?

1909 Manhattan
1936 Triborough

1883 Brooklyn   1903 Williamsburg
1931 George Washington
1961 Throgs Neck

1880 1890 1900 1910 1920 1930 1940 1950 1960 1970 1980 1990 2000 2010

Industrial Era | Depression Era | Auto Era

THE THREE MAJOR PERIODS OF BRIDGE CON-
STRUCTION IN NEW YORK ARE EXEMPLIFIED BY THE
BROOKLYN BRIDGE (1883), THE GEORGE WASHING-
TON BRIDGE (1931), AND THE VERRAZANO-NARROWS
BRIDGE (1964)

BRIDGES HAVE INCREASED IN SIZE AND BECOME
RESTRICTED TO AUTOMOBILE USE; THE MOST
RECENT MAJOR BRIDGE CONSTRUCTED IN THE CITY
WAS THE VERRAZANO-NARROWS, IN 1964

LENGTH: SPAN / OVERALL   COST (2010 $)

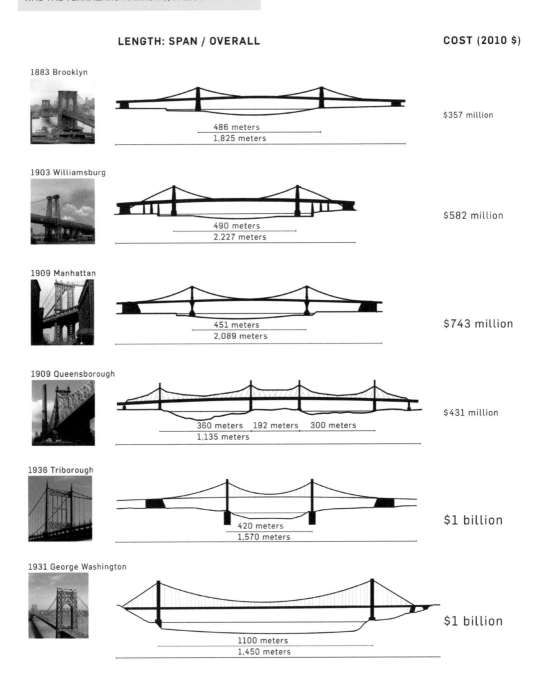

1883 Brooklyn

486 meters
1,825 meters

$357 million

1903 Williamsburg

490 meters
2,227 meters

$582 million

1909 Manhattan

451 meters
2,089 meters

$743 million

1909 Queensborough

360 meters   192 meters   300 meters
1,135 meters

$431 million

1936 Triborough

420 meters
1,570 meters

$1 billion

1931 George Washington

1100 meters
1,450 meters

$1 billion

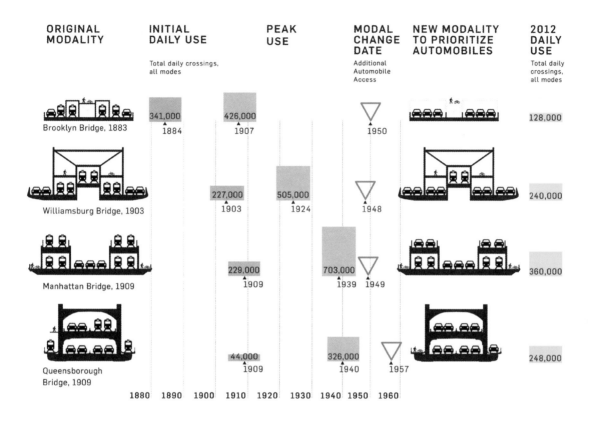

ORIGINAL MODALITY	INITIAL DAILY USE	PEAK USE	MODAL CHANGE DATE	NEW MODALITY TO PRIORITIZE AUTOMOBILES	2012 DAILY USE
	Total daily crossings, all modes		Additional Automobile Access		Total daily crossings, all modes
Brooklyn Bridge, 1883	341,000 1884	426,000 1907	1950		128,000
Williamsburg Bridge, 1903	227,000 1903	505,000 1924	1948		240,000
Manhattan Bridge, 1909	229,000 1909	703,000 1939	1949		360,000
Queensborough Bridge, 1909	44,000 1909	326,000 1940	1957		248,000

1880 1890 1900 1910 1920 1930 1940 1950 1960

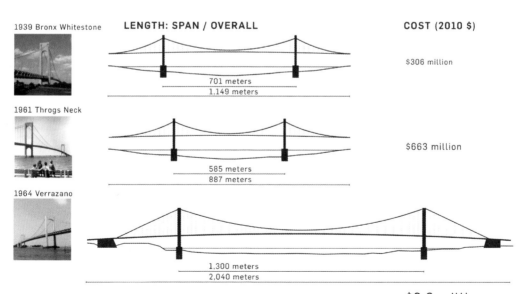

	LENGTH: SPAN / OVERALL	COST (2010 $)
1939 Bronx Whitestone	701 meters / 1,149 meters	$306 million
1961 Throgs Neck	585 meters / 887 meters	$663 million
1964 Verrazano	1,300 meters / 2,040 meters	$2.2 million

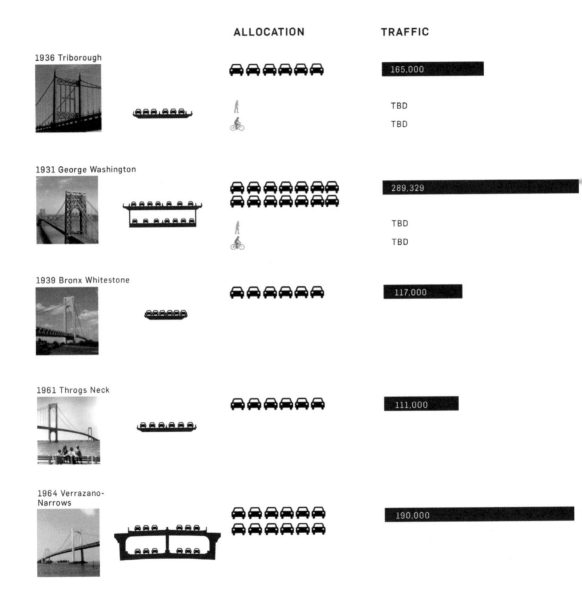

ALLOCATION   TRAFFIC

1936 Triborough

165,000

TBD
TBD

1931 George Washington

289,329

TBD
TBD

1939 Bronx Whitestone

117,000

1961 Throgs Neck

111,000

1964 Verrazano-
Narrows

190,000

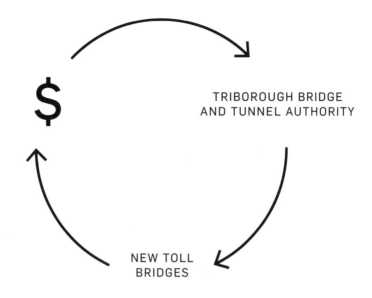

The Moses Profit Loop

TRIBOROUGH BRIDGE
AND TRAVEL AUTHORITY
REVENUE

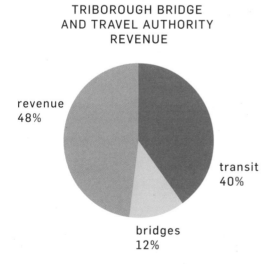

TRIBOROUGH BRIDGE
AND TRAVEL AUTHORITY
EXPENDITURES

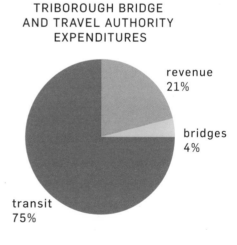

AFTER THE CREATION OF THE
METROPOLITAN TRANSPORTATION
AUTHORITY, BRIDGE TOLLS WERE
USED TO SUBSIDIZE MASS TRANSIT
IN NEW YORK CITY

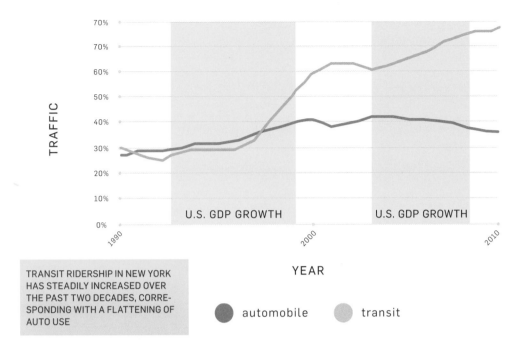

TRAFFIC

YEAR

70%
70%
60%
50%
40%
30%
20%
10%
0%

1990   2000   2010

U.S. GDP GROWTH   U.S. GDP GROWTH

● automobile   ● transit

TRANSIT RIDERSHIP IN NEW YORK
HAS STEADILY INCREASED OVER
THE PAST TWO DECADES, CORRE-
SPONDING WITH A FLATTENING OF
AUTO USE

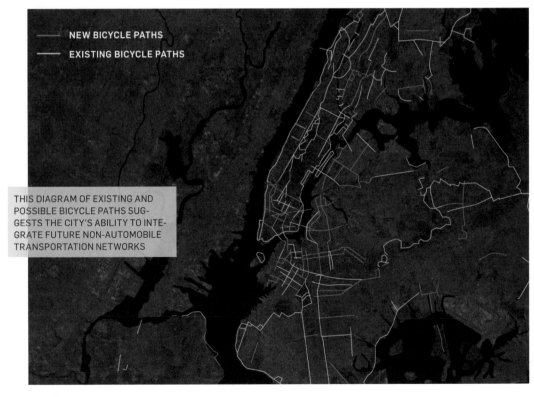

NEW BICYCLE PATHS
EXISTING BICYCLE PATHS

THIS DIAGRAM OF EXISTING AND
POSSIBLE BICYCLE PATHS SUG-
GESTS THE CITY'S ABILITY TO INTE-
GRATE FUTURE NON-AUTOMOBILE
TRANSPORTATION NETWORKS

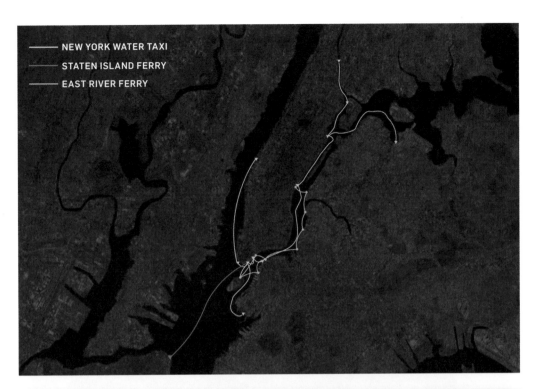

NEW YORK WATER TAXI
STATEN ISLAND FERRY
EAST RIVER FERRY

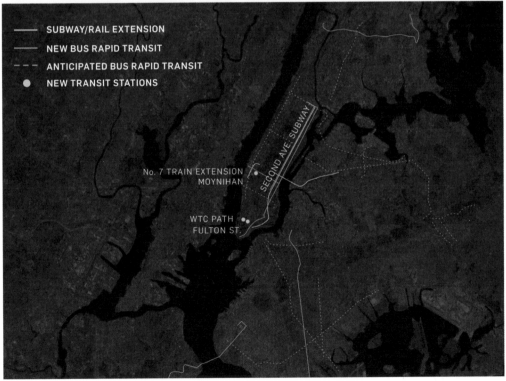

SUBWAY/RAIL EXTENSION
NEW BUS RAPID TRANSIT
ANTICIPATED BUS RAPID TRANSIT
NEW TRANSIT STATIONS

No. 7 TRAIN EXTENSION
MOYNIHAN

SECOND AVE. SUBWAY

WTC PATH
FULTON ST.

NEW YORK HAS ONE OF THE LOWEST PER CAPITA RATES OF ELECTRICITY CONSUMPTION OF ANY MAJOR CITY IN THE UNITED STATES

Chicago 8,143 kWh

New York City 4,696 kWh

San Francisco 6,753 kWh

Phoenix 13,344 kWh

Dallas 16,116 kWh

Houston 14,542 kWh

COAL

NATURAL GAS

PETROLEUM

NUCLEAR

HYDROELECTRIC

BIOMASS

WIND

GEOTHERMAL

SOLAR

PROPOSED POWER PLANT

EXISTING POWER PLANT

TAPPAN ZEE, Tarrytown, N.Y.

E. 42ND ST., New York City, N.Y.

THE LION'S SHARE OF ELECTRICITY CONSUMED BY NEW YORK CITY IS GENERATED BY CITY AND STATE POWER PLANTS

NEW YORK CITY
ENVIRONMENTAL
AND ENERGY
ANALYSIS

ENERGY
ANALYSIS

To better understand the existing environmental landscape of the New York metropolitan area, the studio's environmental and energy research sought first to establish the relative energy consumption of New York in comparison with other major U.S. cities. While New York might not have a reputation for being particularly "green," it is in fact the most efficient among large U.S. cities when evaluated by energy usage per resident. This is in large part due to its density and robust, well-utilized public transportation system. Understanding that there is still room for improvement, the research also investigated the "landscape" of energy consumption over the course of a typical year in New York City. The research concluded that the amount of petroleum-based energy required to power houses, businesses, and industrial buildings is less than the amount of energy required to power petroleum-based methods of transportation. Finally, to establish potential on-site methods of energy production, the research identified wind and tidal power as the most likely sustainable resources for the project sites.

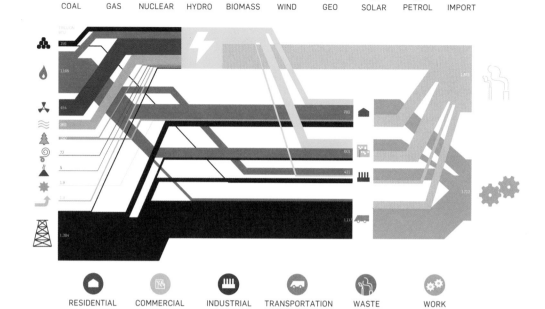

2009 New York State Energy Flow

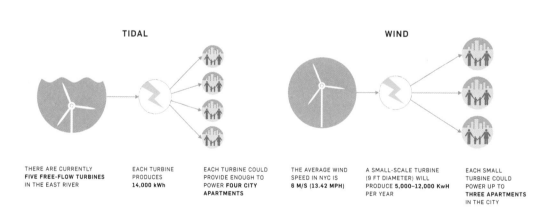

TIDAL

THERE ARE CURRENTLY
FIVE FREE-FLOW TURBINES
IN THE EAST RIVER

EACH TURBINE
PRODUCES
14,000 kWh

EACH TURBINE COULD
PROVIDE ENOUGH TO
POWER **FOUR CITY
APARTMENTS**

WIND

THE AVERAGE WIND
SPEED IN NYC IS
6 M/S (13.42 MPH)

A SMALL-SCALE TURBINE
(9 FT DIAMETER) WILL
PRODUCE **5,000–12,000 KwH**
PER YEAR

EACH SMALL
TURBINE COULD
POWER UP TO
THREE APARTMENTS
IN THE CITY

TIDAL AND WIND ENERGY ARE PARTICU-
LARLY APPLICABLE TO THE EAST RIVER
AND TAPPAN ZEE SITES

ENVIRONMENTAL ANALYSIS

As the Hudson River flows from Lake Champlain to the Atlantic, its momentum is dispersed by the shallow site of the Tappan Zee Bridge, which sits at the river's widest segments. The routine dredging required to keep a navigable 35-foot-deep channel through this portion of the Hudson disrupts the riverbed's ecology and stirs up pollutants. Two peninsulas in the segment—Croton Point and Piermont—help to moderate the disruptions caused by human interventions in this ecosystem. In the estuarine conditions of the Hudson, these peninsulas produce a marshy effect downriver, resulting in markedly different ecological attributes for the north versus the south sides of each peninsula. The upriver side of each peninsula exhibits rocky, beach-like terrain, with a precise line where clear water meets land. Downriver waterfronts are turbid and shallow, with tall marsh grasses blurring the shoreline. This stoppage serves as a wetland.

Spiny Water Flea

Sea Lamprey

Asian Horned Beetle

Mile-a-Minute Weed

Rock Snot

Giant Hogweed

Sudden Oak Death

Hemmorhagic Septicemia

Water Chestnut

Hemlock Wooly Adelgid

Oriental Bittersweet

Zebra Mussel

THE RIVERS ARE
PLAGUED BY ONGOING
POLLUTION AND INVA-
SIVE SPECIES

Air Pollution

Radiation

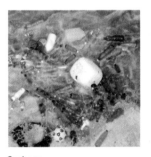

Garbage

Brownfields

Acid Rain

Sewage

Chemicals

Heavy Metals

Salt Marsh + Beach + Side Streets
Jamaica Bay

Industry + Building + Side Streets
**Red Hook / Brooklyn's Navy Yard /
Hunters Point**

Beach + Boardwalk + Side Streets
Coney Island / Rockaway Beach

Esplanade + Buildings + Highway
at Grade
Battery Park City

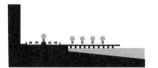

Pier + Esplanade + Highway at Grade
Hudson River Walk

Highway + Park + Side Streets
Riverside Park / Lief Erickson Drive

Highway + Buildings Above
+ Side Streets
Upper East Side

Industry + Highway + Esplanade Above
Brooklyn Heights

Bulkhead + Infrastructure
JFK Airport / LaGuardia Airport

EAST RIVER WATERFRONT TYPES

THE LOWER PART OF THE HUDSON AND EAST RIVERS ARE PARTICULARLY VULNERABLE TO RISING SEA LEVELS AND STORM SURGES

CIVIC & RECREATIONAL

ECOLOGICAL

FOOD & AGRICULTURE

EDUCATION

SCIENCE & RESEARCH

GOVERNMENT

THERE ARE MANY STAKEHOLDERS CONCERNED WITH THE HUDSON WATERSHED

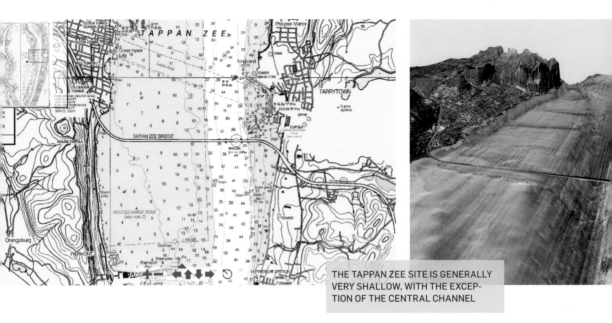

THE TAPPAN ZEE SITE IS GENERALLY VERY SHALLOW, WITH THE EXCEPTION OF THE CENTRAL CHANNEL

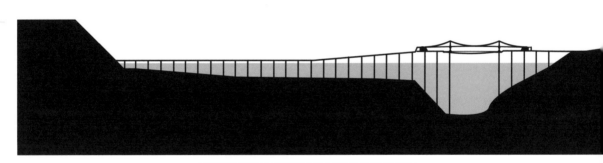

HUDSON RIVER AT THE TAPPAN ZEE BRIDGE SITE

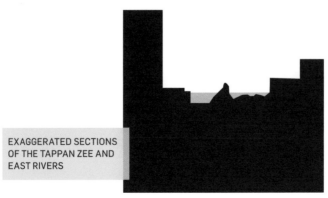

EXAGGERATED SECTIONS OF THE TAPPAN ZEE AND EAST RIVERS

EAST RIVER

THE LOWER PART OF THE HUDSON AND
EAST RIVERS ARE PARTICULARLY VUL-
NERABLE TO RISING SEA LEVELS AND
STORM SURGES

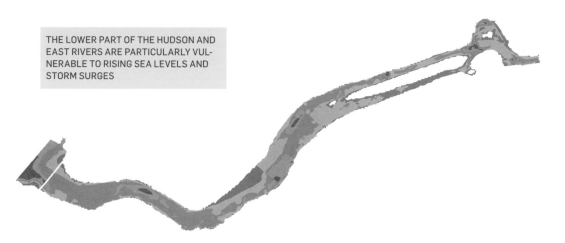

MARKET

RESEARCH

The local market was analyzed for each site, with a focus on the demographic trends and comparable developments currently on the market. The two sites exhibited markedly different characteristics, even from one side of a river to the other. At the 42nd Street site, the Manhattan side showed low population growth, moderate change in housing prices, and relatively low ethnic and household diversity, whereas the Queens side showed high population growth, due to rezoning in Long Island City since 2000; stronger price appreciation, and a high degree of ethnic and household diversity. Nonetheless, the residential developments on either side of the river were similar in terms of their unit mix: a large number of studio, one-bedroom, and two-bedroom apartments were on the market. In contrast, while Tarrytown and Nyack were demographically similar to each other, the market at the Tappan Zee site was different in many ways from the city's. These communities exhibited low population growth and low ethnic diversity, but, unlike the Manhattan and Queens sites, house prices during the 2000s decreased, and new developments at the Tappan Zee cater almost exclusively to families in the form of single-family houses and large apartments, often three bedrooms or more. The larger housing stock reflects the area's current status as a so-called bedroom community.

THERE HAS BEEN SIGNIFICANT POPULATION GROWTH IN LONG ISLAND CITY IN THE PAST DECADE DUE TO REZONING AND THE AVAILABILITY OF LARGE, FORMER INDUSTRIAL PROPERTIES

STABLE
100,000 (2000)
100,000 (2010)

+260%
3,850 (2000)
10,000 (2010)

- over 20% increase
- 10% to 20%
- 0% to 10%
- 0% to -10%
- -10% to -20%
- over 20% decline

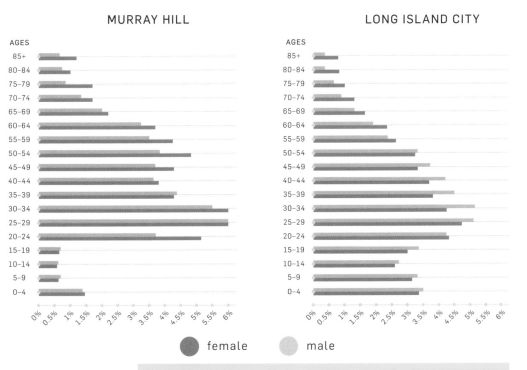

MURRAY HILL

LONG ISLAND CITY

female   male

THE DEMOGRAPHICS ARE MARKEDLY DIFFERENT ON EACH SIDE OF THE RIVER, WITH MORE ETHNIC DIVERSITY AND PEOPLE EMPLOYED IN THE SERVICE SECTOR IN LONG ISLAND CITY

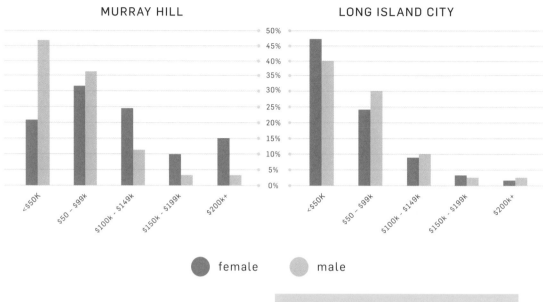

MURRAY HILL

LONG ISLAND CITY

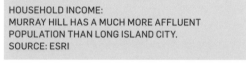

female   male

HOUSEHOLD INCOME:
MURRAY HILL HAS A MUCH MORE AFFLUENT
POPULATION THAN LONG ISLAND CITY.
SOURCE: ESRI

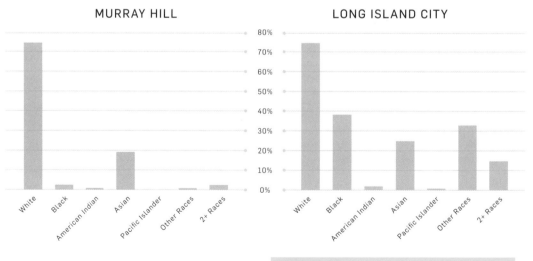

MURRAY HILL

LONG ISLAND CITY

DIVERSITY:
LONG ISLAND CITY IS MUCH MORE ETHNICALLY
DIVERSE THAN MURRAY HILL.
SOURCE: ESRI

MURRAY HILL LONG ISLAND CITY

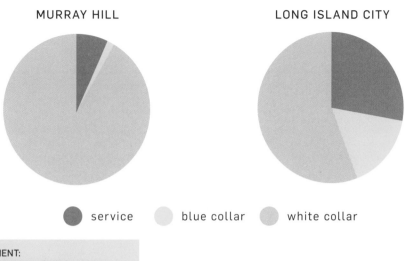

● service   ● blue collar   ● white collar

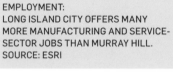

EMPLOYMENT:
LONG ISLAND CITY OFFERS MANY
MORE MANUFACTURING AND SERVICE-
SECTOR JOBS THAN MURRAY HILL.
SOURCE: ESRI

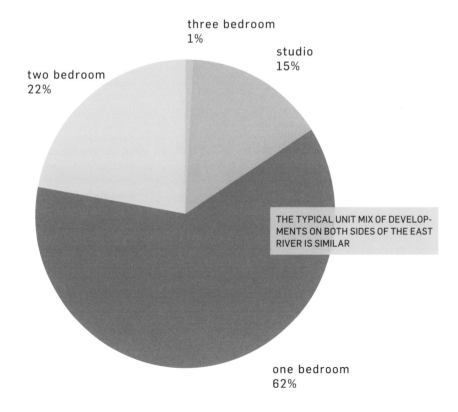

three bedroom
1%

studio
15%

two bedroom
22%

THE TYPICAL UNIT MIX OF DEVELOP-
MENTS ON BOTH SIDES OF THE EAST
RIVER IS SIMILAR

one bedroom
62%

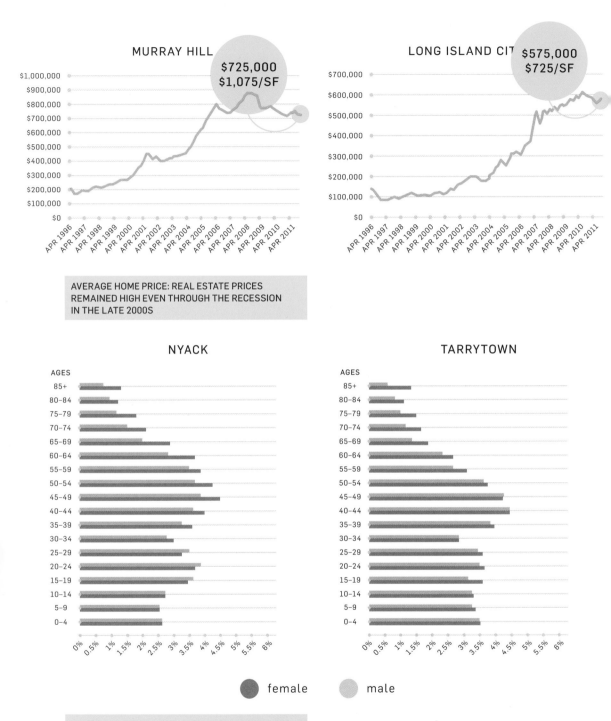

MURRAY HILL

$725,000
$1,075/SF

$1,000,000
$900,000
$800,000
$700,000
$600,000
$500,000
$400,000
$300,000
$200,000
$100,000
$0

APR 1996 APR 1997 APR 1998 APR 1999 APR 2000 APR 2001 APR 2002 APR 2003 APR 2004 APR 2005 APR 2006 APR 2007 APR 2008 APR 2009 APR 2010 APR 2011

LONG ISLAND CIT

$575,000
$725/SF

$700,000
$600,000
$500,000
$400,000
$300,000
$200,000
$100,000
$0

APR 1996 APR 1997 APR 1998 APR 1999 APR 2000 APR 2001 APR 2002 APR 2003 APR 2004 APR 2005 APR 2006 APR 2007 APR 2008 APR 2009 APR 2010 APR 2011

AVERAGE HOME PRICE: REAL ESTATE PRICES
REMAINED HIGH EVEN THROUGH THE RECESSION
IN THE LATE 2000S

NYACK

AGES
85+
80–84
75–79
70–74
65–69
60–64
55–59
50–54
45–49
40–44
35–39
30–34
25–29
20–24
15–19
10–14
5–9
0–4

0% 0.5% 1% 1.5% 2% 2.5% 3% 3.5% 4% 4.5% 5% 5.5% 6%

TARRYTOWN

AGES
85+
80–84
75–79
70–74
65–69
60–64
55–59
50–54
45–49
40–44
35–39
30–34
25–29
20–24
15–19
10–14
5–9
0–4

0% 0.5% 1% 1.5% 2% 2.5% 3% 3.5% 4% 4.5% 5% 5.5% 6%

● female   ● male

NYACK AND TARRYTOWN ARE DEMOGRAPHICALLY
SIMILAR, WITH LARGER HOUSEHOLDS AND MOSTLY
WHITE-COLLAR EMPLOYMENT

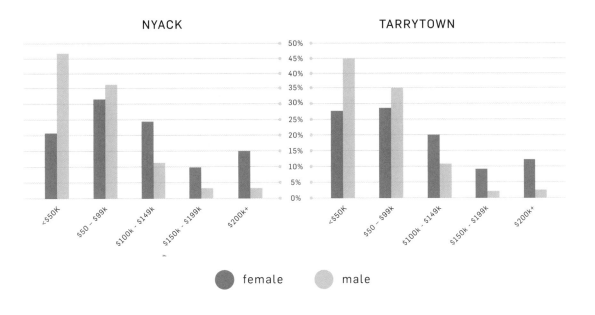

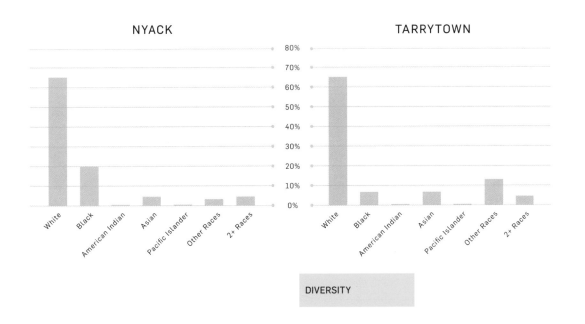

NYACK   TARRYTOWN

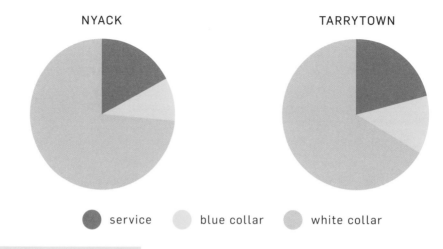

● service   ○ blue collar   ○ white collar

EMPLOYMENT:
RESIDENTS OF NYACK AND TARRY-
TOWN WORK IN SIMILAR BUSINESS
SECTORS. SOURCE: ESRI

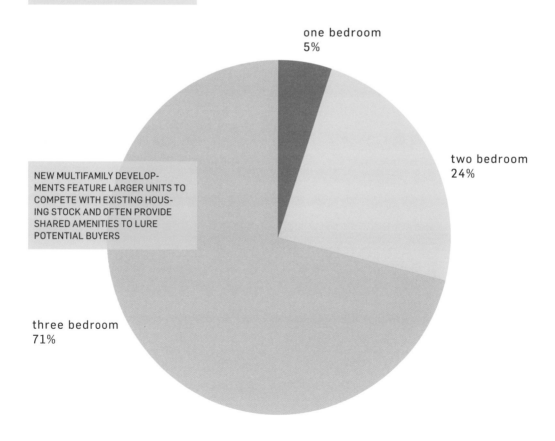

one bedroom
5%

two bedroom
24%

NEW MULTIFAMILY DEVELOP-
MENTS FEATURE LARGER UNITS TO
COMPETE WITH EXISTING HOUS-
ING STOCK AND OFTEN PROVIDE
SHARED AMENITIES TO LURE
POTENTIAL BUYERS

three bedroom
71%

NYACK

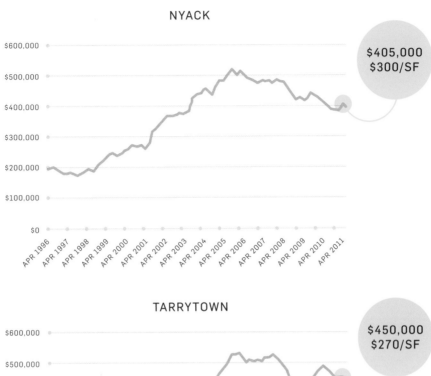

$405,000
$300/SF

TARRYTOWN

$450,000
$270/SF

AVERAGE HOME PRICE:
MUCH MORE AFFORDABLE THAN IN MANHATTAN,
REAL ESTATE IN NYACK AND TARRYTOWN EXPERI-
ENCED A PRICE DECREASE DURING THE RECESSION

● over 20% increase ● 10% to 20% ● 0% to 10%

● 0% to -10% ● -10% to -20% ● over 20% decline

POPULATION GROWTH VARIES
ACROSS THE AREA

PROJECTS

The following pages contain summaries of each group's proposal, including project descriptions, supporting diagrams, visual representations, and model photographs. In addition to meeting the challenge of designing an inhabited bridge and its related social infrastructure, each student group was evaluated on its ability to develop and fully explore the architectural potential of its design strategy based on well-researched, carefully constructed arguments. To demonstrate the viability of the proposed public-private partnership, each group also prepared a pro forma financial projection, which was of considerable importance to the studio but outside the scope of this book.

AERIAL VIEW LOOKING EAST FROM
NYACK TO TARRYTOWN

ENGINEERED REAL ESTATE

JAMES ANDRACHUK, BRYAN KIM,
KARL SCHMECK

The 60-year-old Tappan Zee Bridge needs to be rebuilt. Despite widespread public interest, current plans for the new bridge do not include any mass transit or waterfront access, primarily because the six billion dollar price tag is already a stretch for a public-works project. This proposal tests the role that private development might have in activating the bridge as a livable place in exchange for financing new rail and pedestrian connections. The premise of the project—to create a new land bridge that connects either side of the Hudson at the Tappan Zee Bridge site—emerged from the group's research into the waste materials generated by regional construction projects. By using fill from subway excavation, water-tunnel construction, and harbor dredging in New York City, much of this widest part of the river could be bridged with buildable land, creating ample high-quality, relatively affordable waterfront property and reducing the distance that would need to be spanned by a traditional bridge. The new bridge doubles as real estate, and its form provides the means for funding the project. The project, a collection of waterfront neighborhoods, aims to combine a gradation of urban densities and building typologies to create a vibrant urban environment that offers ease of access, promotes interaction, and naturally appeals to different market sectors. The 540 acres of new land are organized on a modular grid and designed to be walkable communities, connected by central roadways and an integrated transportation hub. Ample amenities both for residents and visitors are strategically located either to create neighborhoods or reinforce the centrality of the transit stations.

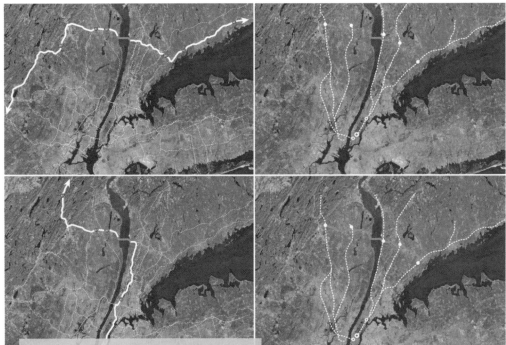

THE TAPPAN ZEE IS IDEALLY SUITED TO CONNECT THE FIVE "FINGERS" OF THE METRO-NORTH RAIL-ROAD, CREATING A CROSS-SUBURB COMMUTER LINE AND BRINGING TOGETHER THE NORTHERN PART OF THE METROPOLITAN REGION. THE BRIDGE COULD PROVIDE HOUSING FOR PEOPLE WHO WORK IN MAN-HATTAN OR IN THE SUBURBS; IT IS ABLE TO SERVE BOTH THE CORE AND PERIPHERY OF THE REGION

THE REGION'S POPU-LATION IS STEADILY INCREASING AND IS EXPECTED TO GROW BY OVER 17% BY 2030

EXISTING CONSTRUCTION PROJECTS IN NEW YORK—THE SECOND AVENUE SUBWAY, WATER TUNNEL NO. 3, AND HARBOR DREDGING, AMONG OTHERS—COULD PROVIDE INEXPENSIVE FILL TO CREATE NEW REAL ESTATE AT THE TAPPAN ZEE BRIDGE SITE

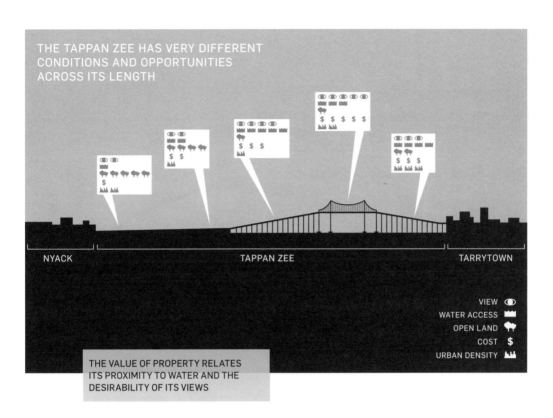

THE TAPPAN ZEE HAS VERY DIFFERENT CONDITIONS AND OPPORTUNITIES ACROSS ITS LENGTH

NYACK

TAPPAN ZEE

TARRYTOWN

VIEW
WATER ACCESS
OPEN LAND
COST
URBAN DENSITY

THE VALUE OF PROPERTY RELATES ITS PROXIMITY TO WATER AND THE DESIRABILITY OF ITS VIEWS

SECTIONAL STUDY MODELS OF INHABITABLE-BRIDGE STRUCTURES

COMPARISON OF MAXIMUM GRADES

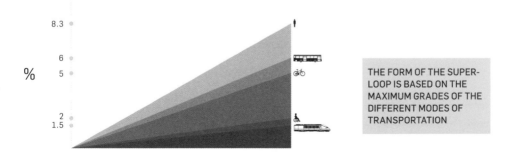

%

8.3

6
5

2
1.5

THE FORM OF THE SUPER-LOOP IS BASED ON THE MAXIMUM GRADES OF THE DIFFERENT MODES OF TRANSPORTATION

Project: Engineered Real Estate

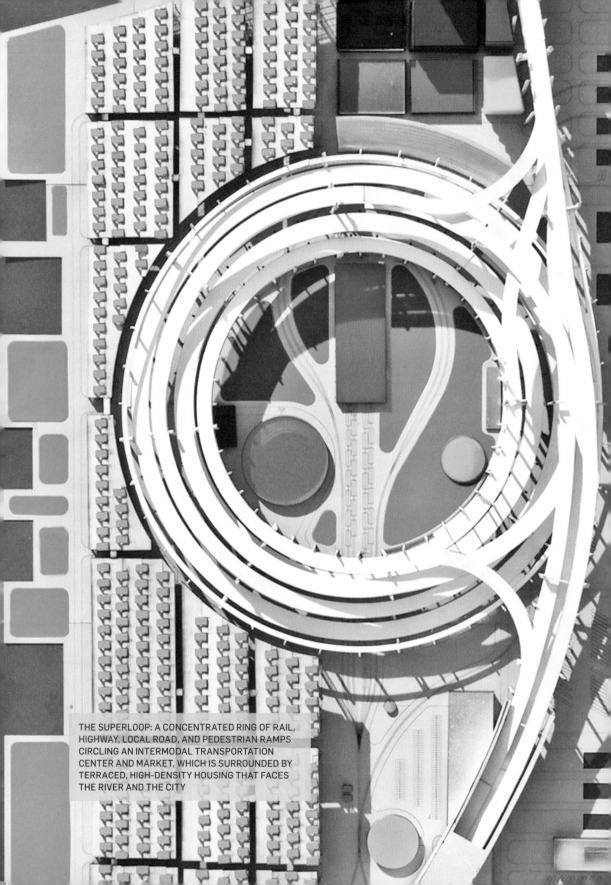

THE SUPERLOOP: A CONCENTRATED RING OF RAIL, HIGHWAY, LOCAL ROAD, AND PEDESTRIAN RAMPS CIRCLING AN INTERMODAL TRANSPORTATION CENTER AND MARKET, WHICH IS SURROUNDED BY TERRACED, HIGH-DENSITY HOUSING THAT FACES THE RIVER AND THE CITY

USING HIGH-SPEED WATER TAXIS, TAPPAN ZEE RESIDENTS COULD HAVE A 30-MINUTE COMMUTE TO MANHATTAN, SIMILAR TO COMMUTE TIMES FOR SUBURBAN QUEENS AND BROOKLYN

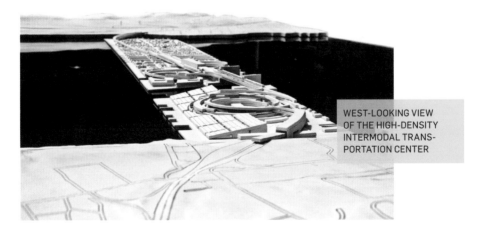

WEST-LOOKING VIEW OF THE HIGH-DENSITY INTERMODAL TRANS-PORTATION CENTER

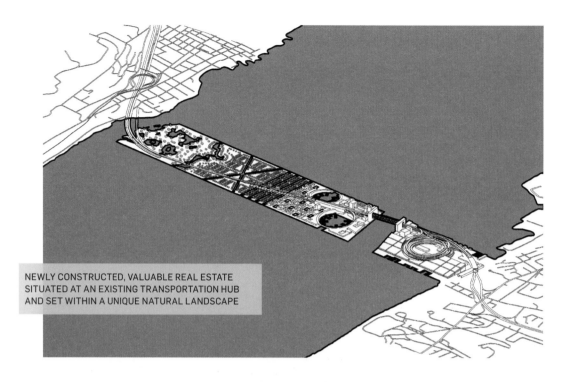

NEWLY CONSTRUCTED, VALUABLE REAL ESTATE
SITUATED AT AN EXISTING TRANSPORTATION HUB
AND SET WITHIN A UNIQUE NATURAL LANDSCAPE

Fill river
with land

Place new highway
on fill

Cut shipping channel

Build highway off-ramps

Relocate Metro-
North stations

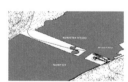

Integrate New York Water
Taxi stop

Establish road grid

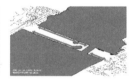

Create waterfront access

Establish zoning plane

Build around pedestrian
amenities

HOW THE BRIDGE
COMES TOGETHER

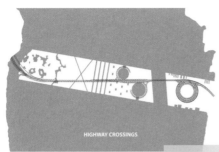

HIGHWAY CROSSINGS

COMPONENTS OF THE BRIDGE

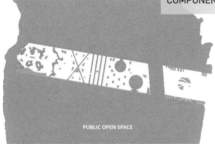

PUBLIC OPEN SPACE

"What if the fill had happened close to Tarrytown? With this particular type of infrastructure—because there is no feedback between the traffic line and the city—the idea of creating real estate to fund the bridge could have happened just as well along the shoreline."

— ALEJANDRO ZAERA-POLO

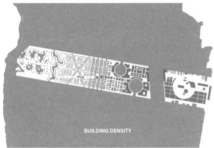

BUILDING DENSITY

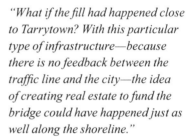

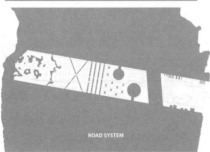

ROAD SYSTEM

"When you presented this, you presented it as sensible to make this gradient a quick pan through all North American housing types across this three-mile bridge. That strange pan, that strange gradient, is itself the takeaway. It's not just a sensible rationale—it's more than the sum of its parts."

— KELLER EASTERLING

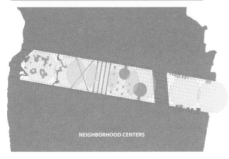

NEIGHBORHOOD CENTERS

Project: Engineered Real Estate

"If we received this master plan from the city, I would think, This is an amazing master plan! And, from here, we would take it to something that would really be wonderful. Right now, it has great potential at a municipal planning level, but, in terms of architecture, lovability, seduction, and exploration, it's not there yet."
— **BJARKE INGELS**

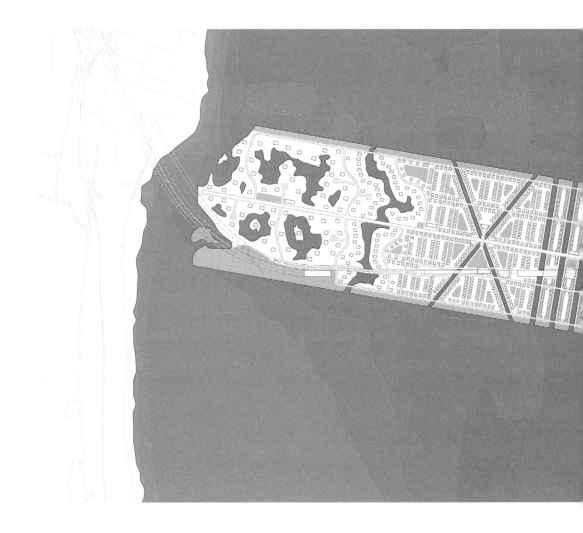

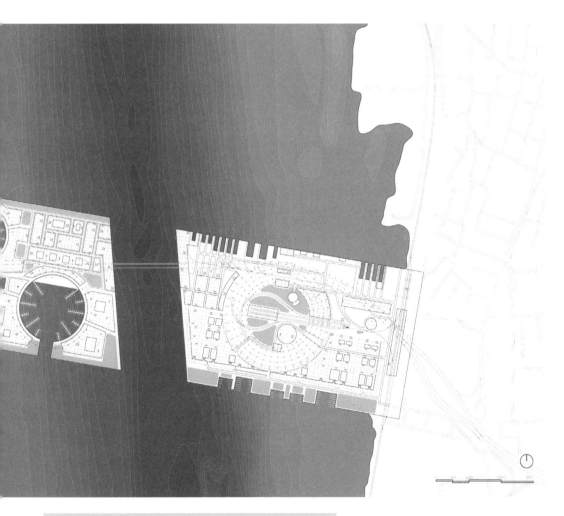

MASTER PLAN OF THE BRIDGE SHOWING A GRADIENT OF DENSITY, FROM VERY LOW ON THE WEST SHORE TO VERY HIGH ON THE EAST SHORE AND CORRESPONDING WITH HIGHWAY, RAIL, AND WATER CONNECTIONS

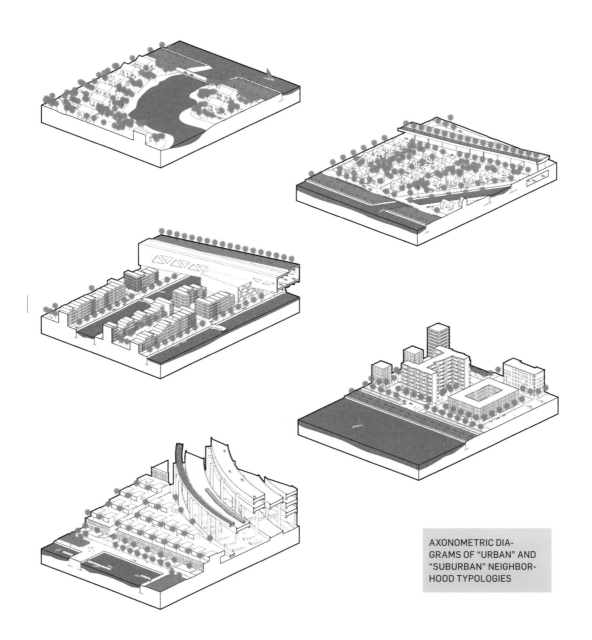

AXONOMETRIC DIA-
GRAMS OF "URBAN" AND
"SUBURBAN" NEIGHBOR-
HOOD TYPOLOGIES

"What I would be looking for is, how are those moments of intersection between the infrastructure and these patterns creating new forms of public spaces or new forms of housing? Because there really is a uniformity to the pattern that we can see in a lot of projects over the past twenty years, using techniques similar to this project. But I feel like you are not using it enough in this process to generate new conditions."
— **GEORGEEN THEODORE**

WINTER VIEW OF CANAL NEIGHBORHOOD, WHICH FEATURES SINGLE-FAMILY HOUSES

LOW-DENSITY HOUSING ON THE WEST END OF THE BRIDGE SITED ON CONSTRUCTED WETLANDS

Project: Engineered Real Estate

MIDRISE APARTMENTS OCCUPY VALUABLE REAL ESTATE AT THE CENTER OF THE BRIDGE, WITH VIEWS OF THE MIDTOWN MANHATTAN SKYLINE TWENTY MILES AWAY

VIEW OF THE LOOP, RIVER, AND LANDSCAPE FROM A TERRACED OUTDOOR SPACE

COMMUTER CENTER: A CIRCULAR MARINA AND
WATER-TAXI STOP IS FORMED BY THE CURVE OF AN
OFF-RAMP AND ENCLOSED BY RETAIL PROGRAM
AND A PEDESTRIAN PROMENADE

STRUCTURE AND ACCRUAL

JAMES SOBCZAK, SUSAN SURFACE,
CRAIG WOEHRLE

The group focused on the management of natural riparian processes to gradually form a new land-scape closely tied to constructed elements. The design aims to minimize the future maintenance costs of the new Tappan Zee Bridge, increase the available waterfront real estate, and provide additional natural amenities to the communities of Tarrytown and Nyack. This new waterfront real estate could be accrued through the carefully managed accumulation of sedimentary deposits along the footings and caissons of the existing bridge. Over a short period of time, this accrued sediment would provide the necessary land upon which to build a new roadway leading to the takeoff point for the new Tappan Zee Bridge. While still maintaining the existing shipping-channel width, the new cable-stay bridge would have much less exposed steel than the existing bridge. These design properties, when combined, greatly reduce the bridge's up-front construction cost as well as its future upkeep. New rail lines and modern stations would also integrate into the design, greatly expanding the commuting possibilities of residents in Tarrytown, Nyack, and their adjacent communities. Once the new bridge is completed, additional land would be allowed to accrue over time along the southern edge of the project, eventually providing new tidal wetland habitats, nature preserves, and an eco-tourism park. To the north of the project, habitable piers and a marina could be constructed to provide real estate opportunities unlike anything else along the banks of the Hudson. The water inlets between these piers, besides sustaining the nature preserve downstream, also could be configured to draw power from daily tidal shifts, thus making the new properties completely energy-independent of Tarrytown and Nyack. All of these elements combine to make a scalable design proposal that can be calibrated to address the area's ever-changing economic and environmental needs.

THE EXISTING BRIDGE AND OPEN
SPACE BEYOND

WINTER VIEW: HOUSING IN THE FOREGROUND, THE MAIN SPAN OF THE BRIDGE IN THE BACKGROUND

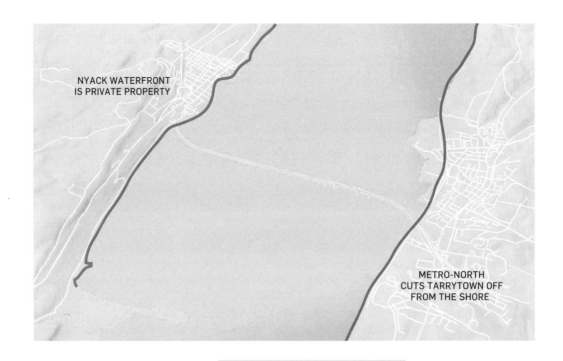

NYACK WATERFRONT
IS PRIVATE PROPERTY

METRO-NORTH
CUTS TARRYTOWN OFF
FROM THE SHORE

THE WATERFRONT IS CUT OFF FROM
THE ADJACENT TOWNS BY INFRA-
STRUCTURE AND PRIVATE PROPERTY

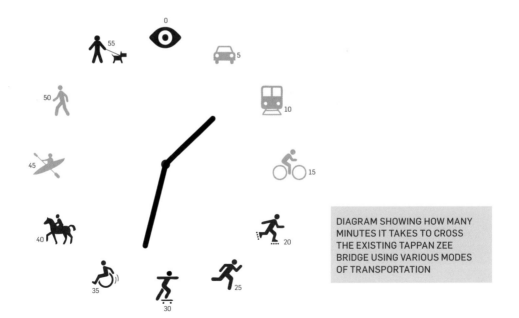

DIAGRAM SHOWING HOW MANY
MINUTES IT TAKES TO CROSS
THE EXISTING TAPPAN ZEE
BRIDGE USING VARIOUS MODES
OF TRANSPORTATION

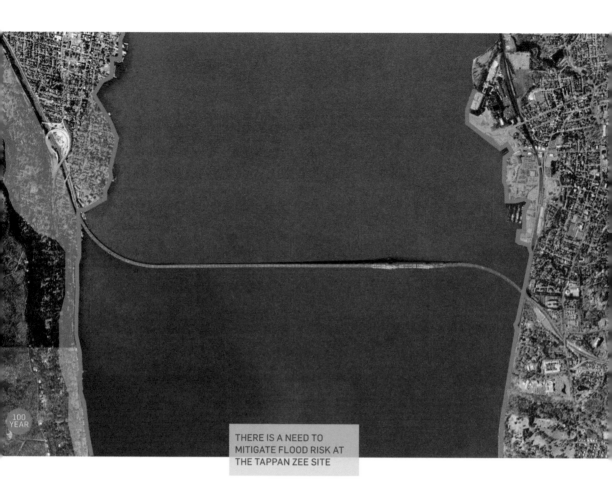

100
YEAR

THERE IS A NEED TO
MITIGATE FLOOD RISK AT
THE TAPPAN ZEE SITE

"It's a foregone conclusion that the bridge is going to be built, so your role is to minimize the risk for the capital that's being put in. It's an interesting zone to play in as an architect. It's not about how much more you put in; it's how little you have to put in to make it convincing."
— JEFFREY INABA

Project: Structure and Accrual

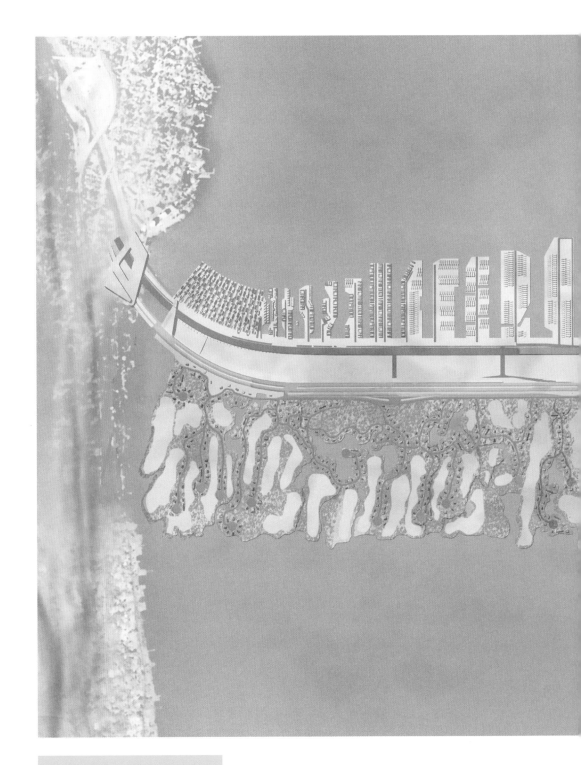

SITE PLAN AT MAXIMUM BUILD-OUT

Project: Structure and Accrual

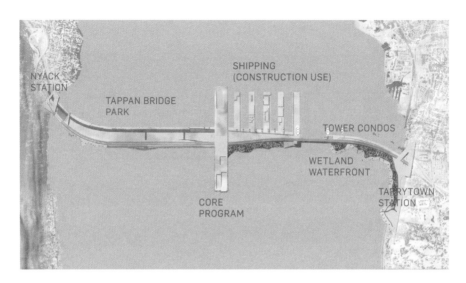

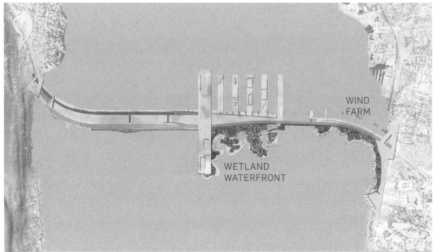

SITE PLAN SHOWING THE PROJECT'S
DEVELOPMENT AT VARIOUS PHASES

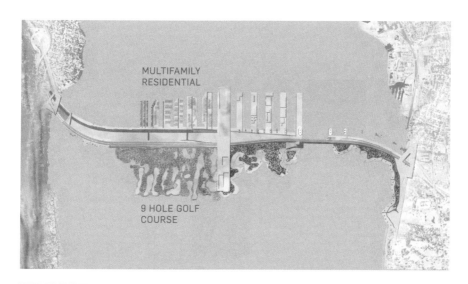

MULTIFAMILY
RESIDENTIAL

9 HOLE GOLF
COURSE

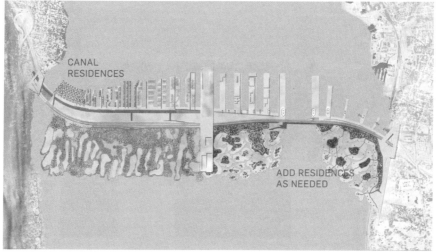

CANAL
RESIDENCES

ADD RESIDENCES
AS NEEDED

*"The great insight you had is that you could
build rapid transit because of the way the
infrastructure is designed. But is the land use
optimized when you're using isolated piers that
segregate the land use? It might work better if
there were larger, continuous parcels."*
— **NANCY PACKES**

ORDER OF OPERATIONS: CONSTRUCT GROUND; PRESERVE WATERFRONT AND SHIPPING CHANNEL; PRESERVE OLD SPAN, CONSTRUCT PARALLEL SPAN; SPLIT PROGRAM ALONG NORTH-SOUTH "EQUATOR"; STORMWATER CONTROL TO SOUTH; BRIDGE RECLAMATION DATUM; NEW CAISSONS TO THE EAST; TRANSIT-PROGRAMMATIC HUBS AT INTERSECTIONS; FORM DETERMINED BY PROGRAM, CONSTRUCTABILITY, AND FINANCING

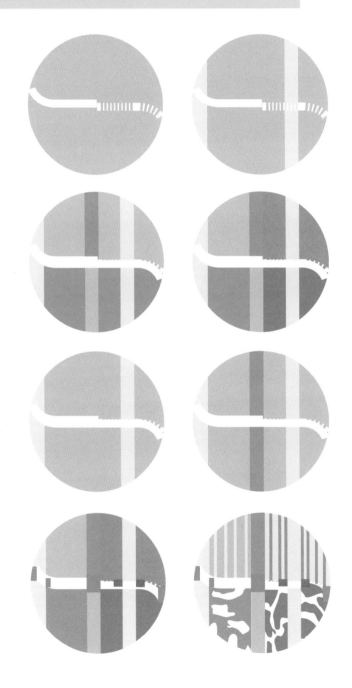

NEIGHBORHOOD OF SINGLE-FAMILY
HOMES SITUATED ALONG THE CANAL

VIEW OF THE CENTRAL AXIS-PARK

WETLANDS AND GOLF COURSE

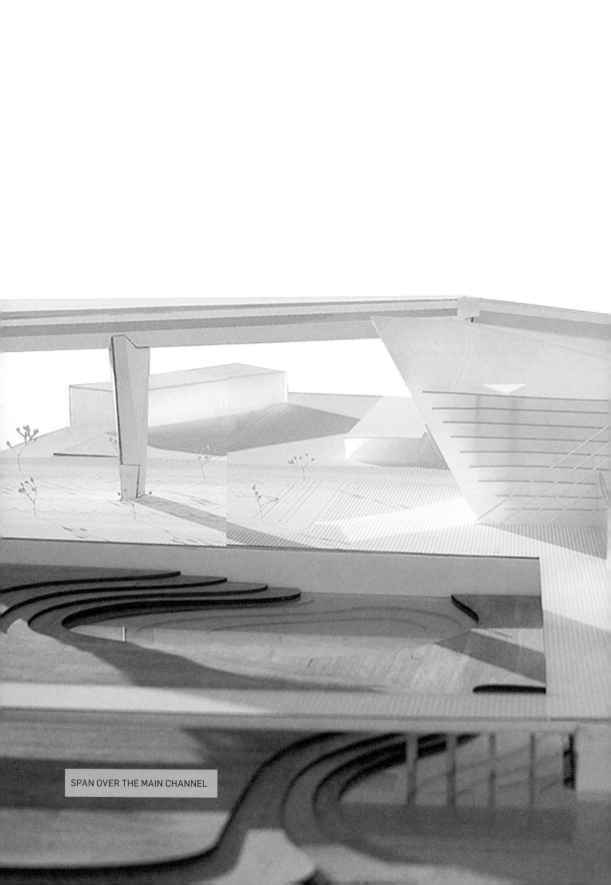

SPAN OVER THE MAIN CHANNEL

"Sometimes, in architecture, there's a discrepancy between quirky-charming and feasible-rational-conventional, and what I like about the project is that there is bravery toward being romantic."
— **BJARKE INGELS**

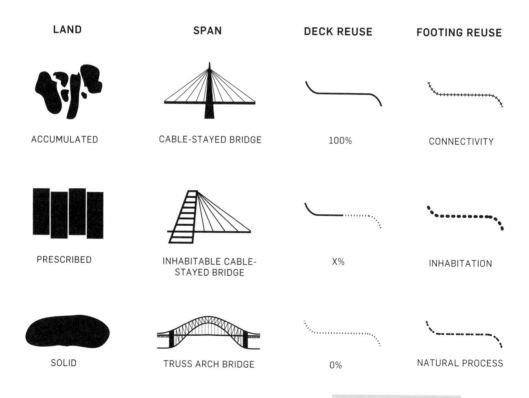

LAND	SPAN	DECK REUSE	FOOTING REUSE
ACCUMULATED	CABLE-STAYED BRIDGE	100%	CONNECTIVITY
PRESCRIBED	INHABITABLE CABLE-STAYED BRIDGE	X%	INHABITATION
SOLID	TRUSS ARCH BRIDGE	0%	NATURAL PROCESS

STRUCTURAL STRATEGIES

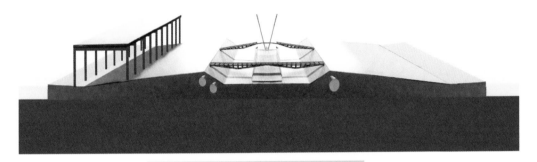

"TRANSIT BERM," SHOWING RAIL, PEDESTRIAN,
AND HIGHWAY INFRASTRUCTURE

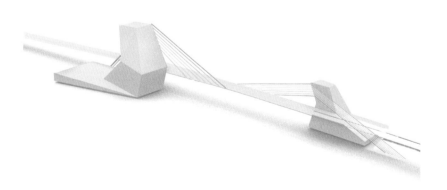

THREE-DIMENSIONAL STUDY OF THE CABLE-STAYED
SPAN ACROSS THE MAIN CHANNEL OF THE HUDSON

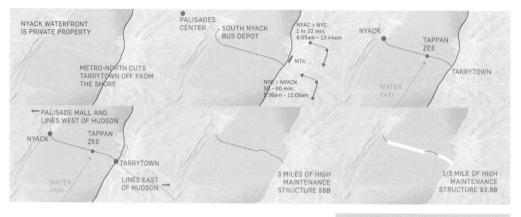

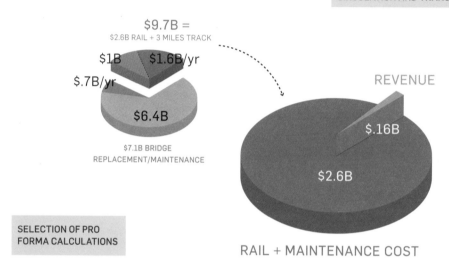

$9.7B =
$2.6B RAIL + 3 MILES TRACK

$1B $1.6B/yr

$.7B/yr

$6.4B

$7.1B BRIDGE
REPLACEMENT/MAINTENANCE

REVENUE

$.16B

$2.6B

SELECTION OF PRO
FORMA CALCULATIONS

RAIL + MAINTENANCE COST

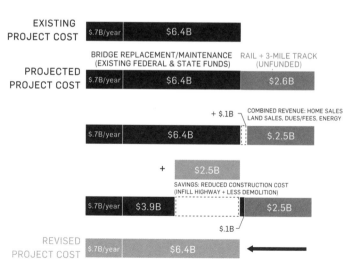

EXISTING
PROJECT COST

$.7B/year $6.4B

BRIDGE REPLACEMENT/MAINTENANCE   RAIL + 3-MILE TRACK
(EXISTING FEDERAL & STATE FUNDS)   (UNFUNDED)

PROJECTED
PROJECT COST

$.7B/year $6.4B $2.6B

+ $.1B   COMBINED REVENUE: HOME SALES
         LAND SALES, DUES/FEES, ENERGY

$.7B/year $6.4B $2.5B

+ $2.5B

SAVINGS: REDUCED CONSTRUCTION COST
(INFILL HIGHWAY + LESS DEMOLITION)

$.7B/year $3.9B $2.5B

$.1B

REVISED
PROJECT COST

$.7B/year $6.4B

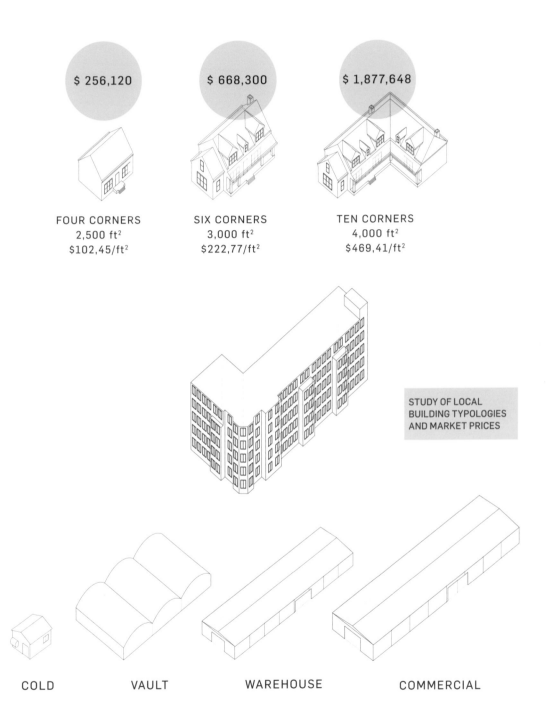

$ 256,120

$ 668,300

$ 1,877,648

FOUR CORNERS
2,500 ft²
$102,45/ft²

SIX CORNERS
3,000 ft²
$222,77/ft²

TEN CORNERS
4,000 ft²
$469,41/ft²

STUDY OF LOCAL
BUILDING TYPOLOGIES
AND MARKET PRICES

COLD

VAULT

WAREHOUSE

COMMERCIAL

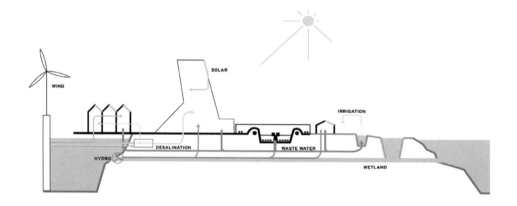

DIAGRAM OF ENERGY-GENERATION STRATEGIES
INTEGRATED WITH THE DESIGN

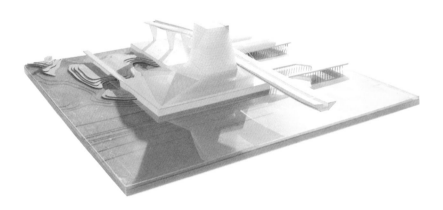

MODEL OF THE MAIN SPAN

THE MAIN PEDESTRIAN PATH, SHOWING GREEN
SPACE, RETAIL, AND TERRACED HOUSING WITH VIEWS
OF THE EAST RIVER AND THE MIDTOWN SKYLINE

EXTENDING
THE CITY

NICKY CHANG, AVI FORMAN,
MARCUS ADDISON HOOKS

The Durst-BIG studio began with a question: Is it possible to fully align private equity and public benefit? Could a public-private partnership go beyond the perfunctory privately-owned public spaces, or POPS, that developers often barter in exchange for a higher floor-area ratio? Could one solution generate a return on investment for the client as well as a substantial addition to the public realm on Manhattan's East Side? The group's proposal begins with the observation that property value in New York is highly correlated to its proximity to active street life: people come to New York to be a part of the buzz of activity. This proposal not only offers pleasant scenic views for a community set on the waterfront but incorporates the DNA of Manhattan, a code that consists of thresholds between private and public spaces: stoops, parks, sidewalks, and green-belt pathways. The group produced more than seventy models during the semester, with each iteration exploring the intersection of condominium and rental developments with a public green way, marrying desirable living accommodations with a varied and full pedestrian experience. The project's thesis always maintained a dual focus, holding that an enjoyable and layered pedestrian experience is not only a gift to the public but is in itself a potential multiplier for property value and return on investment.

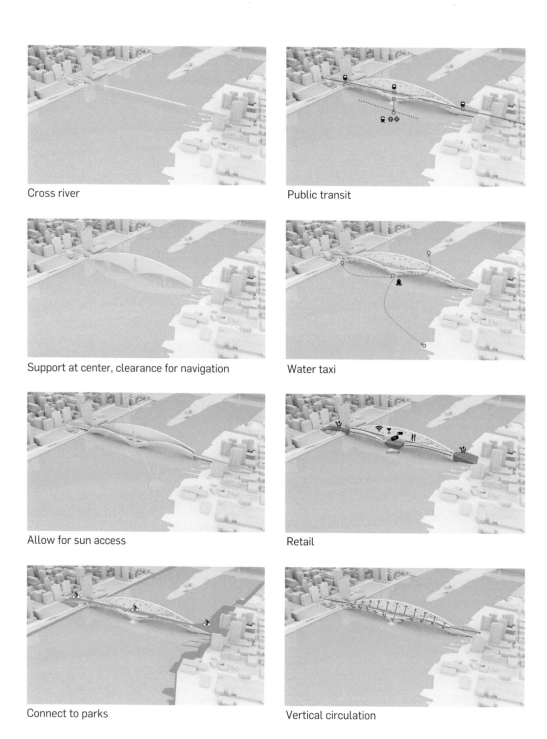

Cross river

Public transit

Support at center, clearance for navigation

Water taxi

Allow for sun access

Retail

Connect to parks

Vertical circulation

ORDER OF OPERATIONS

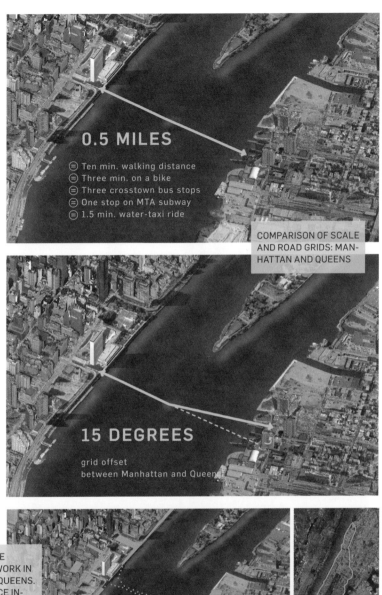

0.5 MILES

= Ten min. walking distance
= Three min. on a bike
= Three crosstown bus stops
= One stop on MTA subway
= 1.5 min. water-taxi ride

COMPARISON OF SCALE AND ROAD GRIDS: MAN-HATTAN AND QUEENS

15 DEGREES

grid offset
between Manhattan and Queens

THE BRIDGE CONNECTS THE EXISTING GREENWAY NETWORK IN MANHATTAN TO PARKS IN QUEENS. THE ACCESS TO OPEN SPACE IN-CREASES REAL ESTATE VALUE. AS OF 2010, CENTRAL PARK IS WORTH $4,579 PER SQUARE FOOT

THE VIEW FROM GANTRY PARK
ON HUNTERS POINT, LONG ISLAND
CITY, QUEENS

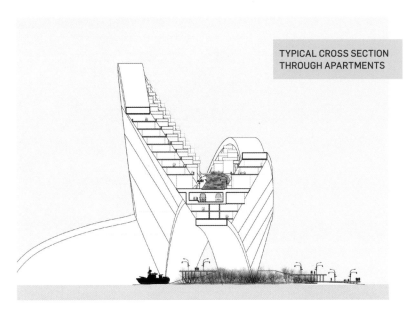

TYPICAL CROSS SECTION
THROUGH APARTMENTS

TYPICAL UNIT INTERIOR

PUBLIC TRANSIT AT THE QUEENS
BRIDGE TOUCHDOWN

AXONOMETRIC CROSS SECTION, SHOWING HOUSING,
RETAIL, AND TRANSPORTATION INTEGRATED INTO
THE STRUCTURE OF THE SPAN

*"There are some empty spots, an abstract grid with a protocol
for how things could be next to each other. Each of those things
could be different. It needs another set of combinants, so that it
becomes quite irregular—lots of openings—creating different
valuations with outdoor spaces."*
— KELLER EASTERLING

COMMERCIAL PROME-
NADE ON THE MANHAT-
TAN SIDE OF THE BRIDGE

THE PEDESTRIAN EXPERIENCE

PUBLIC TRANSIT STATION

TOUCHDOWN OF BRIDGE ON THE MANHATTAN SIDE

"There's such clear reasoning in the way that you're doing things that some of the elements become overarticulated for no reason. You say you take advantage of the space between the top and bottom chords, but the top two are oversized and the bottom one is undersized. The expression of them is greater than they need to be, if you are being rational. And that's where image becomes important."
— **JEFFREY INABA**

One Bedroom Condos
No. of units total: **500**
Average area: **900 sqft**

Two Bedroom Condos
No. of units total: **400**
Average area: **1,050 sqft**

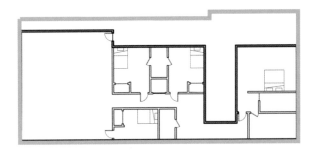

Three Bedroom Condos
No. of units total: **200**
Average area: **1,400 sqft**

Penthouses [4+ bedrooms]
No. of units total: **25**
Average area: **2,200+ sqft**

TYPICAL UNIT PLANS

1/3 RENTALS

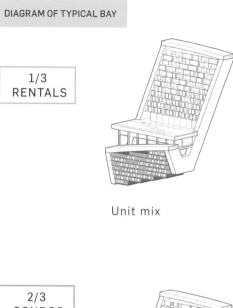

Unit mix

200'-0" CENTER

Circulation-structure integration

2/3 CONDOS

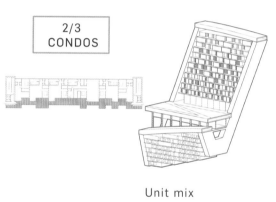

Unit mix

PUBLIC PARK + PRIVATE TERRACES

Open space diagram

12 BAYS

TOTAL SQFT 1.1M

COMMERCIAL + RESIDENTIAL LOBBY

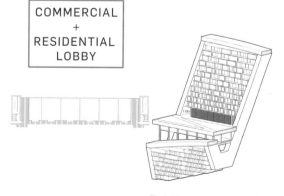

Public-commercial location

"It's a sambadrome *of New York, a carnaval experience. You create all these views, a celebration from one side to another. It's no longer something horrifying that you have to bike over—it gives you the joy of going from one side to another."*
— **JENS HOLM**

"It's great that you were able to get into such design issues and that you were focused on public spaces. It seems a little steep to begin with, but it's nice to have the two intermezzos between the street and the paths. It maybe could have been more urban at the middle. Right now, it's very green, which is nice, but I imagine maybe more of a street."
— **THOMAS CHRISTOFFERSEN**

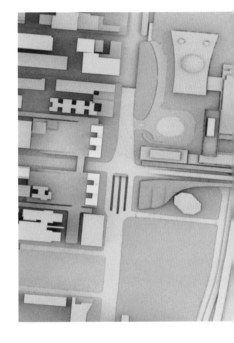

THE DESIGN TAKES ADVANTAGE OF NEW YORK'S "STOOP CULTURE" TO ENCOURAGE STREET LIFE ALONG THE BRIDGE; THIS STRATEGY HELPS WITH URBAN FITNESS, AS ACCESS TO OPEN SPACE PROMOTES HEALTH. SIXTY PERCENT OF ADULTS IN NEW YORK STATE ARE OVERWEIGHT OR OBESE

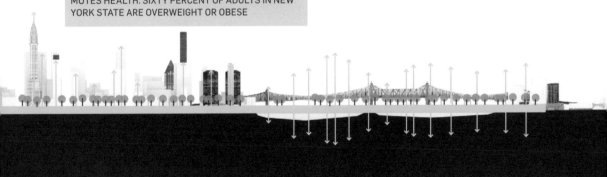

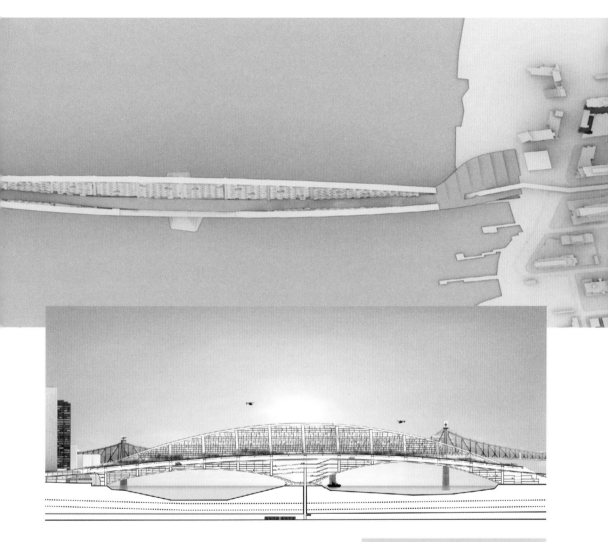

PLAN AND SECTION OF THE BRIDGE,
AT CITY SCALE

Project: Extending the City

VIEW TOWARD QUEENS FROM THE
CENTER OF THE BRIDGE

MODEL, SHOWING CONCOURSE LEVEL

THE GRID, THE BRIDGE, AND THE TOWER

DAVID BENCH, TOM FRYER

New York City grew beyond its beginnings in lower Manhattan using three methodological types: the grid, the bridge, and the tower. The Commissioners' Plan of 1811 created the street grid above Houston Street. Later, starting with the High Bridge in 1848, a series of bridges was constructed to connect the island with the outer boroughs. With horizontal expansion having reached its limit, the tower took center stage, and the skyline began to take on a dynamic role in development. This inhabited bridge connects 42nd Street with Long Island City, and the negotiation of the two street grids across the East River results in a spine of towers that runs along the high point between two navigable channels in the waterway. The creation of a huge new footprint below the bridge serves as an alternative location for the new Cornell technology campus, anchoring the project financially to a large, long-term institutional tenant, with planned future expansion; the new footprint also serves as a link between the financial institutions on Manhattan with Long Island City's plethora of commercial loft and warehouse space, which would be well suited for potential technology start-ups. The towers built on top of the bridge structure are aligned with the Manhattan grid and appear to be a natural extension of the city. As there is no zoning in the middle of the river, these towers are able to achieve considerable heights and densities, relieving pressure for residential construction in Manhattan neighborhoods while not casting shadows on their neighbors. The inhabited bridge is an ideal place for a tower development. While the city's overlay of highway infrastructure divided neighborhoods and disrupted traditional flows of transit, this bridge augments pedestrian, bicycle, and bus routes between the boroughs, serving as a model for integrated future development of greater density, sustainable transit, housing, and institutional growth.

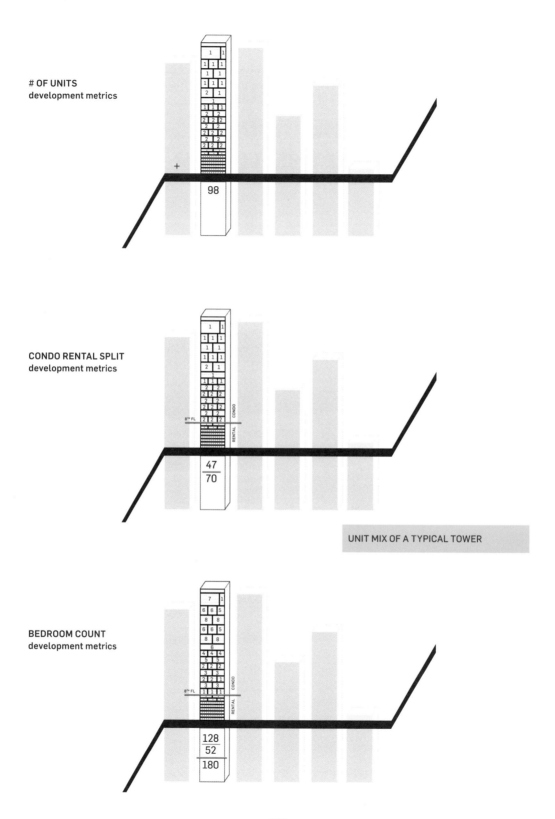

 # OF UNITS
development metrics

CONDO RENTAL SPLIT
development metrics

UNIT MIX OF A TYPICAL TOWER

BEDROOM COUNT
development metrics

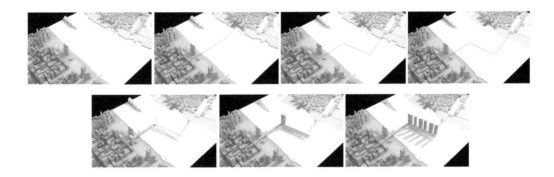

ORDER OF OPERATIONS: EXTEND GRIDS OF MANHAT-
TAN AND QUEENS; CONNECT 42ND STREET TO HUNT-
ERS POINT; MOVE MEDICAL CAMPUS PROGRAM TO
NEW ISLAND; BUILD RESIDENTIAL TOWERS ABOVE

PHASE I
development phasing

PHASE II
development phasing

PHASE III
development phasing

PHASE IV
development phasing

FOUR PHASES OF DEVELOPMENT

Project: The Grid, the Bridge and the Tower

BRIDGE TOUCHDOWN IN THE CENTER
OF THE RIVER

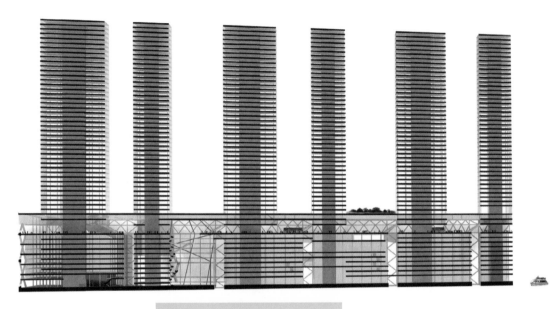

CROSS SECTIONS THROUGH PRIVATE
AND PUBLIC SPACES

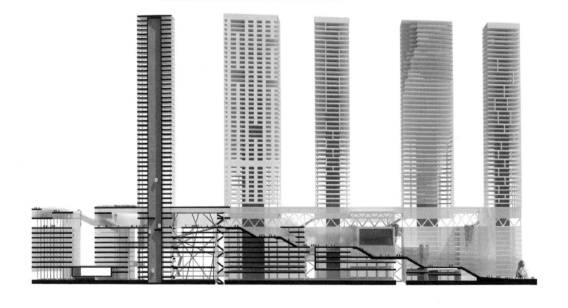

MODEL, SHOWING ELEVATION OF TOWER PODIUM

AERIAL VIEW, FACING NORTH

"I would've considered lying those towers on their side—the whole bridge. That's what Raymond Hood proposed in 1929. It would've been a totally different scheme on the same zigzag armature. It would have released Cornell from its burden of supporting the towers or building these towers on top of occupied labs, which need stability."
— **ROBERT A. M. STERN**

VIEW OF THE BRIDGE
FROM THE FRANKLIN D.
ROOSEVELT MEMORIAL
ON ROOSEVELT ISLAND

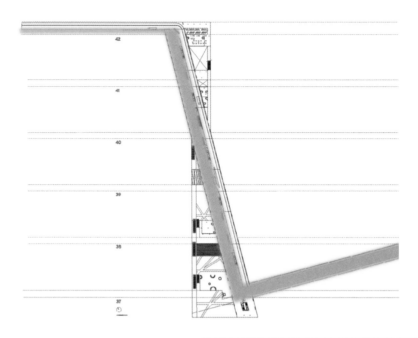

TYPICAL PLANS: ROADWAY AND GREENWAY

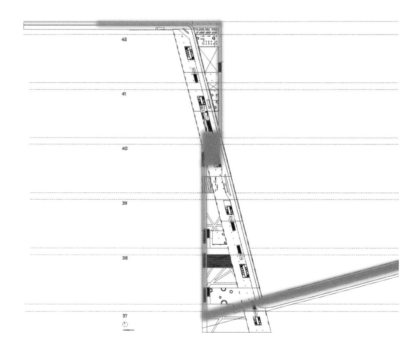

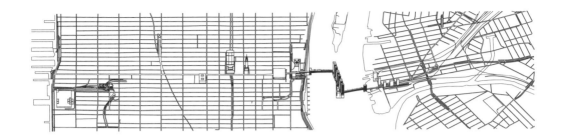

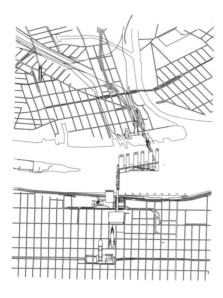

PLANS AT THE CITY SCALE, SHOWING THE BRIDGE'S POSITION BETWEEN MANHATTAN AND QUEENS

"There seems to be an internal struggle with how to merge transportation with the built environment. The cool aspect of mixing the bridge with living is, how to make it one thing rather than two. The bridge doesn't really become part of the tissue of the thing. So, it becomes, in essence, Roosevelt Island again, minus the cool feature of floating between the buildings. There's a general struggle in the studio to merge the two things."
— **JENS HOLM**

THE INTERIOR OF THE BRIDGE, SHOW-
ING PEDESTRIAN WALKWAY, BICYCLE
PATHS, AND PUBLIC TRANSIT

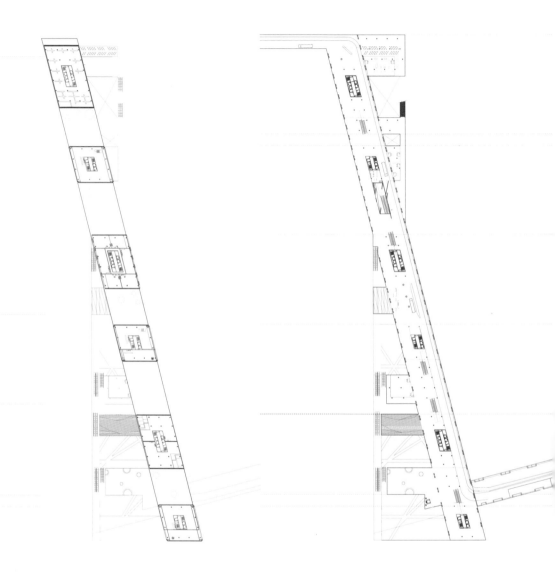

PLANS OF MAJOR FLOORS WITHIN THE CENTRAL
PART OF THE BRIDGE AND TOWERS

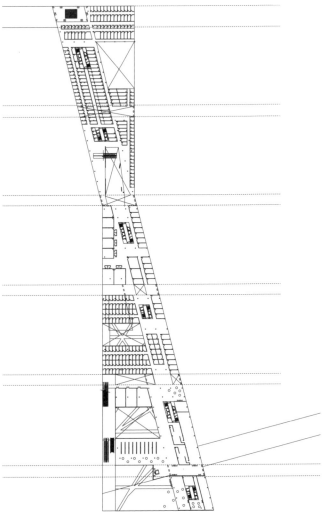

"To make this work, it has to be a very special, exclusive place. You tried to put so much in there, it's lost its sense of what it is. If you just had the first phase of it, with something very special—you talked about developing that space on the water—that's where people would want to be. But you have almost nothing there. What is the space? And why would people want to be there?"

— DOUGLAS DURST

SHIFTING PROGRAM FROM ADJACENT ISLANDS
ONTO THE BRIDGE ITSELF

VIEW FROM HUNTERS POINT AND THE MOUTH OF NEWTOWN CREEK

VIEW OF MIDTOWN MANHATTAN FROM A UNIT INTERIOR

APPROACH FROM THE MANHATTAN SIDE

179
Project: The Grid, the Bridge and the Tower

MODEL, SHOWING BRIDGE LOOKING EAST TOWARD
QUEENS FROM MANHATTAN

REVIEWS

THOMAS CHRISTOFFERSEN AND BJARKE INGELS

IMAGE CREDITS